Sex and death in protozoa

SEX AND DEATH IN PROTOZOA

The history of an obsession

GRAHAM BELL

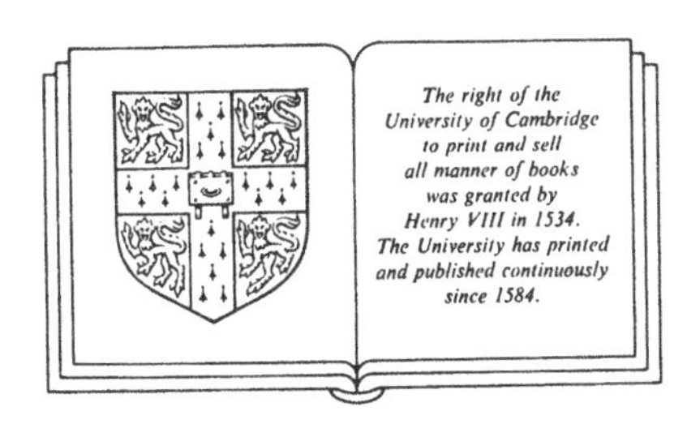

CAMBRIDGE UNIVERSITY PRESS

Cambridge

New York New Rochelle

Melbourne Sydney

CAMBRIDGE UNIVERSITY PRESS
Cambridge, New York, Melbourne, Madrid, Cape Town, Singapore, São Paulo

Cambridge University Press
The Edinburgh Building, Cambridge CB2 8RU, UK

Published in the United States of America by Cambridge University Press, New York

www.cambridge.org
Information on this title: www.cambridge.org/9780521361415

First published 1988
This digitally printed version 2008

A catalogue record for this publication is available from the British Library

Library of Congress Cataloguing in Publication data

Bell, Graham, 1949–
Sex and death in protozoa: the history of an obsession/Graham Bell.
p. cm.
Bibliography: p.
Includes index.
1. Protozoa–Reproduction. 2. Protozoa–Mortality. 3. Protozoa–Genetics.
I. Title.
QL366.B45 1989
593.1′0415–dc19

ISBN 978-0-521-36141-5 hardback
ISBN 978-0-521-05670-0 paperback

For my mother

CONTENTS

LIST OF ILLUSTRATIONS

LIST OF TABLES

PREFACE

This book began when, out of curiosity, I glanced at the contents of the first issue of the Quarterly Review of Biology and saw an article with the irresistible title of 'Eleven thousand generations of *Paramecium*'. Its author, L. L. Woodruff, was describing how he had maintained a strain of *Paramecium aurelia* in isolate culture for 19 years, with no sign of any decrease in vitality. My first thought was that this would provide an excellent illustration of the futility of research unguided by adequate theoretical understanding. On second thoughts, there might be more to the problem than met the eye; and little by little I was drawn into reading most of a huge literature, the bulk of it written between 1890 and 1920, generated by the speculation that sex has some ill-defined rejuvenating property that made it indispensable to the indefinite prolongation of a line of protozoans. Although most geneticists and protozoologists (I am neither) are aware of this literature, it has long since faded from the foreground, and the musty volumes in which it lies entombed are nowadays little read. I suspect that most people have been put off by the almost mystical properties ascribed to sex by many of these early workers, at a time when Mendelism had only just been rediscovered and the concept of the genotype was still novel. The attempts to 'revivify' failing cultures with extract of pancreas, for example, read rather strangely today. A large part of this book is simply historical, therefore, born of my fascination with the attempts of these early geneticists to grapple with ideas of sexuality and immortality, and of my own attempts to understand their results in modern terms. But this in turn led me to think more deeply about the limits to the fidelity with which genetic information can be transmitted through a very long sequence of generations, and this more general question forms the second facet of the book. I have tried to report my findings so that they can be understood by any moderately well-informed reader, from the undergraduate level upwards; to the specialists who will be irritated by the necessary shallowness of many sections I offer my apologies in advance.

This project was conceived, and most of the data retrieval and analysis completed, at McGill, where the Biology Department continues to offer an enormously stimulating intellectual milieu. It was supported in part by funds from the Natural Sciences and Engineering Research Council of Canada. Most of the text and a good part of the theory was done at the University of Sussex, who provided unstinted hospitality during my sabbatical leave. In particular, I have benefited more than I can readily express from conversations with John Maynard Smith. I only hope that he will forgive me for any ideas of his that he finds embedded without acknowledgment in the text; they will be the best ones.

Graham Bell
Montreal
November 1987

1

The question of protozoan immortality

Birth and death are notoriously the only two certainties in life. Men have always desired immortality, have been willing to pay any price in order to procure it, and have always been denied it. They have not ceased to ask questions about it. Since the origins of scientific biology, two recurrent themes have been whether all organisms are born, and whether all must die; and the notion that there is a deep connection between these two invariants has fuelled speculation from the earliest times down to the present day. Spontaneous generation was gradually excised from the scientific curriculum, or at any rate pushed far back into the geological fog, but the possibility of immortality has not been so easy to dispose of. All our familiar animals and plants must surely die, as we must ourselves; even when protected against all the usual rigours of life, the approach of death is eventually signalled by a progressive deterioration, an irreversible process of senescence. It was not obvious that the same should be true of the new world of minute creatures revealed by the invention of the microscope towards the end of the seventeenth century. Protozoans like the *Amoeba* and *Paramecium* familiar from introductory biology classes grow, divide into two apparently identical cells, grow and divide again. They may readily be killed, of course, by almost any sort of minor physical or chemical insult, but in principle, given favourable conditions and protected from shocks, perhaps the cycle of growth and division would continue indefinitely, and thereby prove that natural death was not after all inevitable. Ehrenberg (1838) appears to have been the first to publish this speculation, and by doing so ushered in one of the great debates of biology.

Ehrenberg's suggestion was opposed by Engelmann (1862, 1876) and Butschli (1876), who found that after a long time in culture ciliate protozoans begin to show morphological abnormalities which they interpreted as senescent degeneration akin to that of metazoan cells. At the same time, it was shown that during sexual conjugation in these hypotrichs – first correctly interpreted by these authors – many organelles were resorbed and later regenerated. They proposed, therefore, that protozoans were mortal just as metazoans were, but that, like the metazoan egg, they were rejuvenated by sex. Engelmann put this very strongly: 'The demonstration of this process of reorganization, of rejuvenescence, makes it, as it seems to me, entirely unnecessary to search for any other effect of conjugation' (Engelmann, 1876). This 'rejuvenescence', in the subsequent development of the theory, might occur in either of two ways: by a chemical stimulation caused by the mixing of two dissimilar protoplasts, in a manner analogous to the activation of eggs by artificial parthenogenesis, a subject brought into great vogue at about this time by Jacques Loeb; or by a change in the 'fundamental organization' of the products of conjugation, just as embryogenesis normally requires contributions from two parents.

These views were memorably opposed by August Weismann (1889, 1891; see Kirkwood and Cremer, 1982), on the basis of his ideas about the independence of the germ plasm from the soma. Weismann argued that any organism is bound to suffer a series of minor accidents whose cumulative effect will at last be seriously weakening. Senescence evolves as a device to rid the species of worn out, reproductively valueless individuals and allow them to be replaced by their younger and more vigorous progeny. Since the immortality of somatic cells would be useless or even damaging, selection would not hinder the evolution of specialized cell types, which being unable to replicate themselves, were bound to die. Indeed, senescence would be of immediate advantage because materials and energy diverted from the increasingly expensive task of maintaining aging somatic cells could be more profitably deployed to increase reproductive output. In these ideas we can see the germs of several modern theories, including the 'mutation–accumulation' theory of aging suggested by Edney and Gill (1968), and Williams' concept of life history evolution (Williams, 1957). Germ cells, however, must be an exception, since if they were to age the species would become extinct; they therefore retain a vicarious immortality by virtue of their passage from generation to generation. Now, protozoans – he argued – are all potential germ cells, capable of giving rise to new progeny by fission, and therefore share the immortality of the metazoan germ line. Sex, in Weismann's view, was a quite unrelated phenomenon, having no direct connection with senescence. Its function was rather to provide the variation on which continued evolution by natural selection depends.

However, despite Weismann's attempts to uncouple the problems of sex and death, by the last quarter of the nineteenth century they had become inextricably entangled in a single question, to which an answer would be strenuously sought for the next fifty years: could a line of protozoans be propagated indefinitely by vegetative fission, or was eventual extinction certain in the absence of sex? Some early investigations by Dujardin (1841) and Balbiani (1882) seemed to show a decline of vitality – the rate of vegetative fission – in aging cultures, but the generality of senescent decline and its reversal by sexual conjugation was not firmly established until the classical work by Maupas (1888, 1889). In his very extensive experiments he not only laid the technical and conceptual foundations for almost all subsequent work on the subject, but claimed to demonstrate that protozoan cultures pass through a life-cycle comparable to that of metazoans, except, of course, that it is the life-cycle of a clone rather than of an individual. Cultures may be maintained for several hundred generations, but eventually morphological degeneration, often a decrease in size, begins to set in. The animals begin to look shrunken and distorted; the rate of feeding drops; some structures may disappear entirely; even the micronucleus and macronucleus may dwindle and be lost. At last, worn out and enfeebled, the culture dies out. Just like an individual metazoan, the lineage descending from a single protozoan passes through the stages of youth, maturity, old age and inevitable death. These stages are most characteristically distinguished by the ability to undergo sexual conjugation, which is low in youth, greatest in maturity, and disappears in old age; Maupas himself neither consistently observed nor attributed much importance to the senescent decline in the rate of fission that was to be described by so many later authors. During maturity, conjugation arrested the process of senescence, restored youthful vigour to the cell, and thus made continued existence possible. Maupas' experiments, therefore, seemed to establish for sex the fundamental biological role of erasing in each generation the stains and creases of old age.

At about the same time, an independent series of observations was being made by Richard Hertwig (Hertwig 1903), who attempted to subsume his own and Maupas' results under a single general theory. As the life-cycle progresses, the volume of the nucleus increases relative to that of the cytoplasm, supposedly because the nucleus continually extracts and sequesters material from the cytoplasm. This leads eventually to a physiological imbalance which impairs function and induces senescence. Senescence is prevented, therefore, by any process which by reducing the volume of the nucleus restores an optimal 'nucleoplasmic ratio'. This can be achieved, for instance, by the fragmentation of the ciliate macronucleus and the resorption of some of these fragments by the cytoplasm. Such processes rarely restore the nucleus precisely to its proper size, however; it usually remains too large, and the optimal nucleoplasmic ratio must be recreated by further growth of the cytoplasm. (There is some discrepancy here between Maupas and Hertwig: Maupas' animals senesced by getting smaller, while Hertwig usually observed an increase of size in aging cultures.) Only sexual union was completely effective in restoring vigour, perhaps because the introduction of foreign material into the cell somehow slowed down the reaction between nucleus and cytoplasm. Hertwig did not claim that fertilization increased vegetative vigour; on the contrary, its immediate consequence would be to reduce the rate of fission until the foreign nuclear material became adjusted to the new cytoplasm in which it found itself.

Hertwig elaborated these ideas into a general theory of the life-cycle. For example, male gametes have a nucleus which is very large relative to the cytoplasm, while the reverse is true for female gametes; sexual fusion is therefore necessary to restore the optimal nucleoplasmic ratio. He was an influential figure, whose followers (e.g. Gerassimoff 1901, 1902; Popoff 1908, 1909; Rautmann 1909) gave rise to a large literature. During the 1920s, however, publications became fewer, and eventually work on the nucleoplasmic ratio ceased. It was instead Maupas' influence which continued to be felt, with the mainstream of research being devoted to long-term cultural studies.

Ironically, the great majority of these studies concerned themselves with a proposition repeatedly and explicitly denied by both Maupas and Hertwig–Butschli's contention that the leading characteristic of senescence in protozoan cultures was a gradual decline in the rate of fission, which could be reversed by sexual conjugation. Fabre-Domergue (1889) pointed out early in the debate that both a decline in the fission rate and Maupas' senile morphological changes could be caused by a slow deterioration in culture conditions. Joukowsky (1898) agreed, and managed to maintain *Paramecium* through some 450 generations, though his cultures eventually declined and died. The most vocal opponent of the life cycle concept was Enriques (1903, 1905), who took *Glaucoma* through nearly 700 generations without conjugation and with no decline in the fission rate. He attributed his success to frequent changes of medium; cultures kept in medium which was often allowed to grow stale declined and died in the usual Maupasian manner. Impressed by the tedium of transferring isolate lines to fresh medium every day, Enriques concluded that what becomes exhausted in most experiments is not the vitality of the organisms but the patience of the investigator.

It was at about this time that two rival schools became established in North America, under G.N. Calkins and his pupil L.L. Woodruff. Calkins and his collaborators worked on a number of ciliates and invariably found a limited Maupasian life cycle culminating in death, the continued existence of the organisms being effected only by sexuality. Woodruff retorted by raising his famous line of *Paramecium aurelia* for over ten thousand generations with no signs of flagging vitality and without conjugation at any time. Between 1902 and 1940 these two schools, together with a number of European workers, generated most of the voluminous literature I have abstracted and analysed below. Some of this work was rather shoddy, even by the standards of the time, and contributed only to the confusion which continued to surround the main issue. Much of it, however, was meticulous and patient research carried out over longer periods of time than most granting agencies would nowadays permit, and the passage of time has not reduced its value nor blunted its relevance. The most extensive and exact experiments concerning the crucial relationship between vitality and conjugation were published at the close of this period by H.S. Jennings. Jennings' work, however, ended the tradition; he died shortly afterwards, and left no successors. A little earlier, he and his pupil T.M. Sonneborn had discovered the existence of mating types in *Paramecium*, and it was this topic, rather than the fusty old Maupasian controversy, that the new generation of protozoologists elected to follow up. Texts on protozoology published in the first quarter of this century (e.g. Calkins 1926) devote large sections to cultural studies of vitality, and Jennings' 1929 review of protozoan genetics centres around this topic; Wichterman's 1953 text on *Paramecium* relegates the topic to a short chapter late in the book; recent texts (e.g. Nanney 1980) mention the subject only in passing, and in Jones' recent monograph on ciliates (Jones 1974) neither 'senescence' nor 'vitality' appear in the index, and Calkins, Woodruff and Jennings are entirely omitted from the bibliography. The problem was never resolved; after fifty years it was simply abandoned. In this essay I propose to retell this old and tangled tale, and in the retelling to show how the careful and long-continued observations of these forgotten protozoologists do after all yield a crucial clue to the vexatious problem of the function of sex.

2

Sex and reproduction in ciliates and others

Ciliates were the favourite subjects for isolation culture, since being relatively large – the larger species are easily visible to the naked eye – and freeswimming they are readily observed and manipulated. Before describing the behaviour of the cultures, I shall first briefly describe ciliate reproduction and sexuality, to introduce technical terms that I shall make frequent use of later. More detailed accounts of some crucial processes are deferred to the appropriate sections.

2.1 Reproduction

The ciliates that I shall be describing reproduce exclusively by a transverse binary fission, a single cell dividing at right angles to its long axis to give rise to two daughter cells. Because the animal is not symmetrical about the plane of division, the two fission products are not perfectly identical, and for a short period following division anterior and posterior fragments can be distinguished. These differences, however, soon disappear. A more serious complication is introduced by the fact that ciliates possess two different kinds of nuclei. The micronucleus, or 'generative nucleus', is a small body usually possessing two sets of chromosomes. During fission it divides by a regular mitosis, so that the chromosomes, organised on spindle fibres, are distributed accurately into the two daughter micronuclei. During vegetative growth the micronucleus appears to synthesize little or no RNA, and therefore plays little or no part in normal somatic function; indeed, viable strains which completely lack micronuclei are known from several species. Protein synthesis is thus mediated largely, if not exclusively, by the much larger macronucleus. This is a highly polyploid nucleus, typically comprising the equivalent of 40–80 or even more haploid sets of chromosomes. During cell division the macronucleus does not go through a regular mitosis, but is instead pinched into two roughly equal parts, a process called 'amitosis'. The details of this process are not clearly understood even today. The inequalities that such a crude process inevitably leads to are to a great extent made good by compensatory processes occurring after nuclear division – thus, a new macronucleus with less than the standard complement of chromosomes will go through an extra cycle of DNA replication between cell divisions, while one with too many chromosomes will go through two division cycles with only a single replication. However, there are qualitative consequences of amitosis that cannot be so easily overcome, as we shall see later.

2.2 Sexuality

We are apt to be misled by our familiarity with large animals and plants into confusing sexuality with quite different phenomena such as gender and reproduction. Sex is by no means necessarily associated with any of its usual concomitants, including gametogenesis, the differentiation of male and female, or even the production of offspring. In ciliates, for example, sex is completely decoupled from reproduction: two cells enter sexual conjugation, and two cells emerge from it. The essential feature of sexuality is a qualitative, rather than a quantitative, change – the production of heritable diversity by the mingling of different genomes. This is brought about by a twofold process, in which the number of chromosomes is first halved by a reductional division of the nucleus (meiosis), and later restored by the fusion of two such reduced nuclei (syngamy, or fertilization). Any process involving meiosis and syngamy can be recognised as sexual. A catalogue of the sometimes bizarre variations that can be played on this basic theme is given by Bell (1982); here, I shall be concerned only to emphasize the distinction between the two fundamentally different categories of sexuality, amphimixis and automixis. Amphimixis is, roughly speaking, cross-fertilization. Haploid nuclei produced by one individual are packaged into gametes, which fuse with the gametes produced by another individual. The result, obviously, is the mingling of genes from two different lines of descent. In automixis, fusion occurs between haploid nuclei which have the same immediate ancestry. A single diploid cell can produce two (or more) haploid pronuclei, which may then fuse together, with or without any intervening cell division. Automictic fusion, therefore, occurs within the same line of descent, and corresponds roughly to self-fertilization.

The distinction between amphimixis and automixis is important because of the different consequences of these two modes of sexuality for genetic heterozygosity. Imagine an individual which is heterozygous at some locus, so that its genotype at this locus can be represented as A1A2. Other individuals in the population may be either heterozygous (A1A2) or homozygous (A1A1 or A2A2). This heterozygous individual will produce both A1 and A2 gametes by meiosis. If it is cross-fertilized, then its progeny may be either homozygous (if an A1 gamete fuses with an A1 gamete from another individual, or an A2 gamete with another A2 gamete) or heterozygous (if an A1 gamete fuses with an A2 gamete, or A2 with A1). A heterozygous individual therefore inevitably produces a certain proportion of homozygous progeny. However, this loss of heterozygosity is balanced by the fact that a homozygous parent produces some heterozygous progeny – thus, an A1A1 parent, producing only A1 gametes, will have both A1A1 and A1A2 progeny. Indeed, the fundamental result of Mendelian genetics is that, provided gamete fusion occurs at random, there will be no change in the expected population frequency of any of the three genotypes at this locus through time. No matter how many alleles are segregating in the population, the frequency of each diploid genotype will in a single generation of random mating attain a frequency which thereafter will not change. This frequency is the Hardy–Weinberg equilibrium frequency of elementary population genetics, and the result is the maintenance of genotypic diversity through time so long as the population remains undisturbed by processes such as selection or sampling error. On the other hand, consider an automictic heterozygote. Unless it has some special mechanism for preventing it, half its gamete fusions will be of like nuclei with like – A1 with A1, or A2 with A2 – and half its progeny will therefore be homozygotes. There is, however, no compensatory way in which an automictic homozygote can produce heterozygous progeny. The consequence is a progressive loss of heterozygosity, and in fact it is easy to show that in a self-fertilizing population half the heterozygosity is lost in every generation. Eventually all is gone, and the population comprises only the two homozygous lines A1A1 and A2A2. In general, therefore, heterozygosity is conserved in amphimicts, but more or less rapidly lost from automictic populations.

2.3 Ciliate sexuality

The early history of the study of sexuality in ciliates is a catalogue of confusion, some of which persisted into the period of cultural studies, and perhaps even to the present day. Sexual union in *Euplotes* was reported by King in 1693, but he seems to have been observing fission, and

Leeuwenhoek was probably the first to see conjugation a few years later. This was correctly interpreted as a sexual process by O.F. Muller, but, in a reversal of King's error, later observers (e.g. Ehrenberg 1838) interpreted conjugation as fission. This was finally corrected by Balbiani (1860), but he in turn described sperm, ova and embryos – on the basis of excellent microscopical observations during which he saw spindle fibres in the dividing micronucleus, rounded granules produced by the breakup of the old macronucleus, and, as a final touch, endoparasitic suctorians. The correct interpretation of his 'testis' and 'ovary' as micronucleus and macronucleus was made by Butschli in 1873, and their behaviour during conjugation was finally worked out by Maupas (1889) and Hertwig (1889). Their observations form the basis of the modern interpretation of conjugation, which, shorn of its details and variations, is as follows.

Under certain conditions, ciliates such as *Paramecium* form large agglutinated masses of individuals, within which the animals form pairs by fusion over a large part of the oral surface, in such a way that homologous structures are superimposed. This fusion lasts for up to 24 hours. Soon after it begins, the micronucleus divides three times to form a series of eight products. The first two divisions are meiotic, as conjectured by Maupas and Hertwig and later confirmed by genetic (Sonneborn 1947) and cytophotometric (Dupy-Blanc 1969) analyses, and their four products are therefore haploid. The third division is a mitosis. The two identical haploid products formed by this final division from one of the four products of meiosis persist, and constitute the gametic pronuclei; all the other nuclei degenerate. The two paired individuals then exchange one of these pronuclei, the migratory pronucleus of each fusing with the stationary pronucleus of its partner, to form a syncaryon which reconstitutes the diploid state. During these events the macronucleus begins to degenerate, and eventually becomes resorbed into the cytoplasm. The syncaryon divides mitotically, one of its products developing subsequently into the new micronucleus and the other into the new macronucleus. Meanwhile, the two partner cells, now known as 'exconjugants', separate and swim away.

The course of sexual conjugation differs in detail between species of ciliates, depending on the number of micronuclei present, the number of divisions of the syncaryon, and so forth. In some forms, for example in the sessile suctorians, it is superficially quite different from the process I have described. Two important features, however, are common to all the forms that I shall discuss below. The first is that the exconjugants are genetically identical, by virtue of the mitotic formation and reciprocal exchange of the pronuclei. Conjugation therefore reverses the normal course of sexuality:

from two initially dissimilar individuals are formed two identical ones. In the early literature this was not recognised, and there are several reports of large and apparently heritable differences in division rate, size and liability to conjugate among the pure lines arising vegetatively from sister exconjugants (especially Calkins and Gregory 1913). This was attributed by later authors (e.g. Jennings 1929) to faults in experimental design, but, as we shall see later, it is now acknowledged that stable phenotypic differences between clones may arise despite the genetic identity of their micronuclei. Naturally, the exconjugants from different pairs will be genetically different, and mixing two sexually compatible clones produces an enormous variety of types from the two originally present.

The second point concerns the behaviour of the macronucleus. Since the macronucleus divides in concert with the cell as a whole during binary fission, the macronuclear genome is conserved, within the limits permitted by amitosis. When conjugation occurs, however, the old macronucleus disappears and is replaced by a new one under specifications provided by the micronucleus. The nuclear dualism of ciliates, therefore, usually involves parallel behaviour by macronuclei and micronuclei. On occasion, this parallelism is imperfect, as I have described above. In particular, the partial degeneration of the macronucleus which sometimes occurs between cell divisions, resulting in the extrusion of blobs of nuclear material into the cytoplasm, was interpreted mistakenly by Diller (1936) as a sexual process, which he called 'hemixis'. This process, and other peculiar macronuclear configurations to which some of the earlier workers attached great significance, are probably all concerned with regulating the quantity of macronuclear DNA and have no sexual connotation.

One of these old observations, however, has been amply borne out by more recent work. During his long-term *Paramecium* culture, Woodruff (Woodruff and Erdmann 1914, Erdmann and Woodruff 1916) described a peculiar process of nuclear reorganization which he called 'endomixis'. This was subsequently discovered in several other ciliates, but its nature and significance for long remained obscure. It seemed to involve both a non-meiotic division of the micronucleus and a partial breaking-up of the macronucleus. The matter was eventually resolved by Diller (1936), who found that two quite different phenomena had been subsumed under the general term of endomixis. The first was hemixis, the expulsion of parts of the macronucleus. The second was an automixis, the union of gametic pronuclei within the same cell, without conjugation. Diller, whose interpretation was confirmed genetically by Sonneborn (1939), found that the first two micronuclear divisions were reductional, and that as in conjugation the two surviving pronuclei were sisters. Their automictic fusion therefore immediately creates homozygosity at all loci in the genome. An automictic fusion which preempts the exchange of micronuclei during conjugation has been given a different name – 'cytogamy' – but has the same genetic consequences.

In *Paramecium*, automixis occurs in more or less freely-moving individuals, but in some ciliates it occurs only within a cyst. This is an important technical point, since by preventing cyst formation, or removing cysts, in isolation cultures it is possible in some forms to eliminate not only amphimixis but also automixis from the history of a clone.

Ciliates, then, provide excellent material for the comparative and experimental study of sexuality. Amphimixis occurs regularly, and can be induced or prevented nearly at will by the manipulation of culture conditions; it may occur between clones and amount to outcrossing, but it can also occur within clones and amount to self-fertilization; and by mating exconjugant lines whose previous history is known any desired degree of inbreeding can be achieved. Moreover, the genetic identity of exconjugants allows cytoplasmic and nuclear effects to be separated with unusual clarity. Automixis occurs in some forms but not in others, and where this takes place in a cyst it can easily be prevented. And finally, the macronuclear genome undergoes some degree of reorganization independently of the micronucleus, allowing us to separate somatic from heritable changes.

2.4 Other protists

Most cultural studies involve ciliates, and I will describe the sexuality of the few other protists which have been raised in isolation culture more briefly.

Actinophrys sol (Heliozoa) normally reproduces by an equal binary fission, but sexual processes occur in encysted individuals (Belar 1922, 1924). The encysted cell divides into two daughter cells, called gamonts. These undergo a normal two-stage meiosis with one product degenerating after each division, to form two haploid cells which then fuse automictically. Occasionally two individuals form a common cyst, in which case gamont fusion is amphimictic. Unfused gamonts may develop parthenogenetically. This animal is thus basically an automictic diplont.

Peranema and *Entosiphon* are flagellated algae belonging to the Euglenophyta, all members of which reproduce solely by equal binary fission: the few early reports of gamete formation have never been confirmed, and as far as is known the group is entirely asexual.

Eudorina elegans is a colonial green alga belonging to the Chlorophycophyta. The 16- or 32-celled coenobia reproduces asexually by autocolony formation, each cell giving rise by repeated mitoses to a complete new colony. Vegetative colonies are haploid, and reproduce sexually by producing gametes mitotically, some cells differentiating into packets of microgametic (sperm) and others into single macrogametes (ova). Meiosis occurs at zygote germination, with three of the four products degenerating so that a single germling emerges from the zygote. The colonies are normally dioecious (any given colony produces either sperm packets or ova), but the clones are homothallic (any given clone produces both male and female colonies). It is thus an amphimictic haplont, but clonal selfing gives progeny identical among themselves and with their parents. I have included it among the protists because there is no division of labour between somatic and germinal tissue.

3

Isolation cultures

3.1 The technique

Mass cultures of protozoans tell us relatively little; division rates are almost impossible to measure and conditions cannot be kept constant. To follow the history of a clone accurately it is essential to study isolated individuals in small volumes of medium; this was Maupas' fundamental technical achievement. Isolation cultures of asexual metazoans are relatively easy to set up: a newborn female is segregated in a chamber, and her offspring removed as they appear, until she dies. This is impossible in protozoans which reproduce by binary fission, simply because there is usually no way of distinguishing between 'parent' and 'offspring'. Isolation cultures of protozoans therefore cannot follow the history of an individual, but are used rather to follow a line of descent. Nor is it possible to follow all of the branches of any such line, since a few dozen generations of binary fission would produce many billions of descendants from a single initial cell; rather, a few cells must be selected in each generation as representatives of their line. The technique which, with unimportant variations, was used by all workers after 1902 was as follows. A culture is maintained until a conjugating pair is found. One of the two products of conjugation is isolated and allowed to divide two or three times, producing four or eight descendants. Four or five of these are in turn isolated, and allowed to divide. The descent of each of these forms a 'line', the four or five lines together constituting a 'series'. At intervals of between one and three days, depending on the assiduity of the investigator, a single individual is reisolated from all the products of fission formed within a line. If any line should become extinct, it is replaced by an individual from another line in the same series. The experiment ends when all the lines become extinct, or when the patience of the investigator is exhausted.

The output of such an experiment is a rate of division through time. Only a few early papers (e.g. Calkins 1903) give the raw data; more often, a histogram of average rates through time is presented. Many of these histograms are reproduced in the review by Jennings (1929). Nor is the way in which this average was calculated always specified, but a common method was first to obtain the average rate of division per day for each line from the number of surviving cells in that line, and then to calculate the grand mean of all lines over a five- or ten-day period.

To detect any systematic change in the innate capacity of the organisms to divide, it is essential that the culture medium remain constant, or at least that it should not systematically deteriorate, and that it should be renewed very frequently. It is only recently, however, that chemically defined growth medium has been devised for ciliates and in the period when most of the major cultural studies were done the usual medium was a sterile infusion of hay, cereal or dessicated lettuce leaves to which a more or less constant quantity of some known bacterium was added. The earliest work is sometimes technically crude, allowing several days to elapse between isolations, leaving medium exposed to the air to accumulate a bacterial flora, and using inadequate sterilization techniques. Calkins (1903), for example, speaks of wiping his slides with 'a clean cloth which is used for no other purpose'!

3.2 Selection within clones

Even if a perfectly constant medium could be devised, the technique of isolation culture may itself bias the results. In the first place, the strains chosen for study are likely to be preadapted to laboratory conditions; many authors tested many strains but persevered only with those which flourished. The second source of bias is more insidious. Suppose that there exists some variation in fission rate among the vegetative descendants of a single individual. The individual chosen at random to continue the line is likely to have had ancestors whose fission rate was greater than average. If the differences in fission rate are transmitted through vegetative reproduction, then such variations will accumulate and tend to elevate the average fission rate of the line. This is by no means merely hypothetical: Jennings *et al.* (1932) give an explicit account of how weak lines were removed from a series by selection. Nowadays we might regard it as futile to attempt to select within a pure line, having been taught that Johannsen's classical experiment (Johannsen 1903) settled the question for good. Earlier workers, especially protozoologists, were not so pessimistic. Jennings (1908, 1913) and Ackert

(1916) found that variation in size and fission rate within pure lines was small and did not respond to selection, while de Garis (1927) showed that both size and fission rate were unaffected by a period of experimentally-induced 'monster' formation. Nevertheless, Calkins and Gregory (1913) reported substantial variation in fission rate between the lines descending from each of the four products of the first two divisions after conjugation, and Middleton (1915) made the apparently conclusive observation that the differences between two lines created by choosing in one line the first and in the other line the last individuals to divide were maintained through a long period of balanced selection in which the first and last individuals to reproduce were selected in alternate generations in both lines. At about the same time, several investigators (e.g. Jennings 1916, Root 1918, Hegner 1919) succeeded in establishing heritable morphological divergence within pure lines of various protozoans by long-continued selection. In 1927, Parker repeated Middleton's work, obtaining positive results for

Figure 1. Selection from fission rate within pure lines. Parker (1927; *Paramecium aurelia*) and Middleton (1915; *Stylonichia pustulata*) both give the mean fission rate per day in their fast and slow lines as 10-day averages; let them be F and S respectively in any 10-day period. I have calculated the divergence between lines as (F − S)/(F + S) for each 10-day period, and then plotted the moving average (5-point for Parker's data, 3-point for Middleton's data) of these values. If the difference between the lines created by selection is heritable, divergence should increase with time, perhaps eventually tending to a limit. This seems to be supported by the plots.

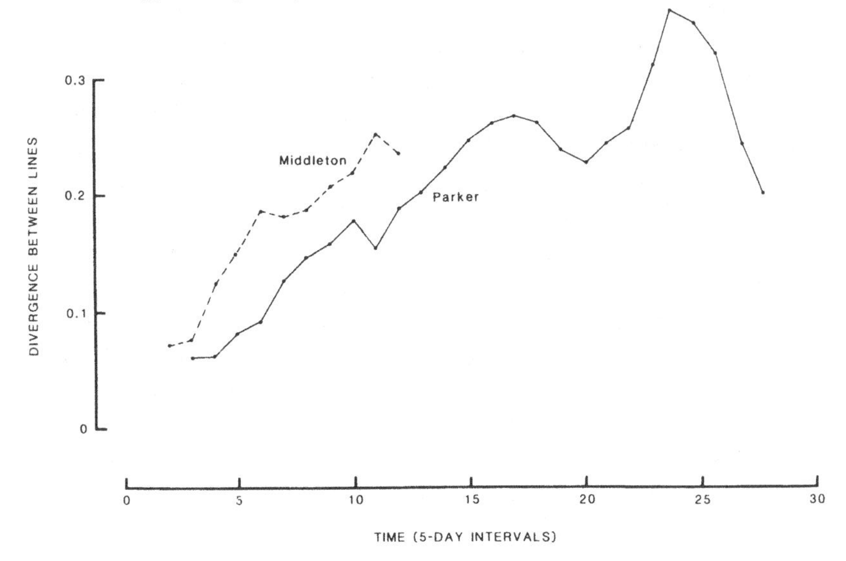

two species, although he failed to procure any heritable response in a third ciliate. Two of Parker's and Middleton's experiments are shown in Figure 1.

These effects might be due to purely somatic differences between the two fission products. In *Paramecium*, for instance, the posterior and anterior fragments can be distinguished for an hour or two after separation, and de Garis (1928) found that selection of the anterior fragment usually (4/5 experiments) generated a higher fission rate (to complicate matters further, Peterson, 1927, reported that although the anterior fragment divides first at 26–30 °C, the posterior fragment divides first at 13–17 °C). de Garis' result is reminiscent of Sonneborn's finding that the posterior halves of a vegetatively reproducing flatworm could be propagated indefinitely, while anterior lines died out (Sonneborn 1930); much later, Siegel (1970) found that the posterior fragment of *Paramecium*, which must regenerate anew the complex oral apparatus, has a greater average fission rate than the anterior fragment. More generally, it is an instance of the much-disputed 'Lansing effect', the decline in vitality of lines descending through many generations of very old mothers. I shall have more to say of this later.

3.3 The genetics of amitosis

It is conceivable, however, that the response to selection observed by Middleton and by Parker represents stable genetic change, despite being evoked within a vegetative line of descent. One obvious possibility is that any asexual lineage will gradually accumulate mutations, some of which will affect the fission rate and can be fixed by selection. Alternatively, the lines which responded may have been automictic. A clone which is initially heterozygous at some locus which affects the fission rate will segregate into two homozygous clones with different rates of fission following the automictic fusion of sister pronuclei, and the nuclear events involved might well pass unnoticed.

What I wish to stress in this section, however, is a third way in which stable variation can arise within strictly asexual lines of descent, which follows from the mechanics of amitosis in the ciliate macronucleus. This nucleus consists of a swarm of genetic elements, which may not be whole chromosomes. Any particular type of element, bearing alleles at some defined locus, may be present in several or many copies. If these elements are diploid, then no immediate problem arises: a heterozygote which expresses an intermediate phenotype will give rise to daughter cells with the same phenotype, since all daughter macronuclei inherit both alleles from their parent. It is well-known, however, that in some ciliates the clone descending from a single heterozygous cell often gives rise to clones which express only one of the parental alleles, so that a uniformly heterozygous stock may give rise to two homozygous clones by strictly vegetative reproduction (for a review of this process in *Tetrahymena*, see Allen and Gibson, 1972). This suggests that the elements which assort during the amitotic division of the macronucleus are haploid – or, if they are diploid, that they are functionally haploid, with one allele suppressed in some copies of the element and its alternative in others. Suppose, for instance, that we consider elements which include the locus A, which may bear either of the two alleles A1 and A2. In the simplest case, only two elements are present in the macronucleus, so that a heterozygote bears one A1 and one A2 element. After replication, there will be two A1 elements and two A2 elements. If these are apportioned at random to the daughter macronuclei, while each daughter receives the same total complement of two elements, then the probability that any given daughter receives two copies of the A1 element is $(2/4)\times(1/3) = (1/6)$; likewise the probability that it will receive two copies of the A2 element is also $1/6$, and thus the probability that a daughter macronucleus will retain the heterozygosity of its parent is only $2/3$. Since this applies with the same force in each generation, the fraction of macronuclei descending from a single heterozygous parent which are themselves heterozygous after t generations of amitosis is only $(2/3)^t$. After a sufficiently long period of time virtually all heterozygosity will have been lost, and the heterozygous stock will have become differentiated into two homozygous clones. It will be appreciated that this loss of heterozygosity applies only to the macronucleus; the micronucleus, which divides mitotically, continues to be heterozygous, and a cell will therefore behave as a heterozygote in sexual reproduction despite expressing a purely homozygous phenotype.

The theory of such processes was worked out by Sewall Wright for the case of plastid inheritance, which is directly analogous to the assortment of haploid elements in amitotic nuclei (see also Schensted 1958). It is a problem in sampling without replacement, and its statistics therefore reflect the properties of the hypergeometric distribution. We are given that the macronucleus at some time t bears M elements of a certain type A, of which M_1 and M_2 are of the alternative states A1 and A2; $M_1+M_2 = M$. After replication there are $2M$ elements, $2M_1$ being A1 and $2M_2 = 2M-2M_1$ being A2. The macronucleus then divides so that each of the two new micronuclei contains M of the A-elements, the two states assorting at random. We want to calculate the probability that the new macronuclei of the succeeding generation $t+1$ will contain any given number x of the A1 elements. This probability is given by

$$\mathrm{Prob}(M_1 = x, t+1) = \binom{2M}{M}^{-1} \sum_{y=x/2}^{(M+x)/2} \binom{2y}{x}\binom{2M-2y}{M-x} \mathrm{Prob}(M_1 = y, t)$$

where the limits of the summation on the right-hand side are integers obtained, when necessary, by rounding down for the upper limit and rounding up for the lower limit. Note that the system has absorbing boundaries at $M_1 = 0$ and $M_1 = M$, since when only one state of the A-element is present all macronuclei must necessarily inherit only that state, mutation being neglected. To emphasize the crucial point that the end-classes at $M_1 = 0$ and $M_1 = M$ must continually increase in frequency we can write these special cases separately as

(*a*) $$\mathrm{Prob}(M_1 = 0,\ t+1) = \mathrm{Prob}(M_1 = 0,\ t) + \binom{2M}{M}^{-1} \sum_{y=1}^{M/2} \binom{2M-2y}{M} \mathrm{Prob}(M_1 = y,\ t),$$

(*b*) $$\mathrm{Prob}(M_1 = M,\ t+1) = \mathrm{Prob}(M_1 = M,\ t) + \binom{2M}{M}^{-1} \sum_{y=M/2}^{M-1} \binom{2y}{M} \mathrm{Prob}(M_1 = y,\ t).$$

A set of diagrams which illustrate this process is given in Figure 2(*a*), showing the derivation of two purely homozygous lines from an initially heterozygous stock during vegetative fission by a process of somatic assortment.

The rate at which this process occurs depends on the number of haploid elements segregating in the macronucleus. The number of A1 elements in daughters descending from a heterozygous macronucleus will be distributed with mean value M_1 and hypergeometric variance $M_1(M-M_1)/(2M-1)$. If we define the frequency of the A1 element as $p = M_1/M$, the variance of the frequency is then $M^2p(1-p)/(2M-1)$. It follows that after sufficient time (about $2M$ generations) the rate of loss of heterozygous macronuclei approaches a value of $1/(2M-1)$ per generation. This has the important consequence that the observed rate of assortment can be used to estimate the number of copies of a haploid element per nucleus; such rates in *Tetrahymena* are commonly about 0.01 per fission (see Table

13 of Allen and Gibson, 1972) and thus suggest the presence of about 50 copies of each element per nucleus.

In developing this argument, it has been assumed that while alternative states of each element assort randomly, the total number of copies per macronucleus is conserved. In the absence of a fairly sophisticated

Figure 2. Somatic assortment. (*a*) The frequency distribution of the number of copies of an allele per nucleus, when the total number of copies of the element is conserved. The original macronucleus is heterozygous and bears $M = 10$ copies of the A element in all, $M_1 = 5$ of type A1 and $M_2 = 5$ of A2. As time goes on, the variance of the frequency distribution of the number of copies of a given type increases, with nuclei accumulating at the two absorbing boundaries $M_1 = 0$ and $M_1 = 10$, it being assumed that each daughter macronucleus receives exactly 10 copies. Eventually, virtually all nuclei contain only A1 or A2.

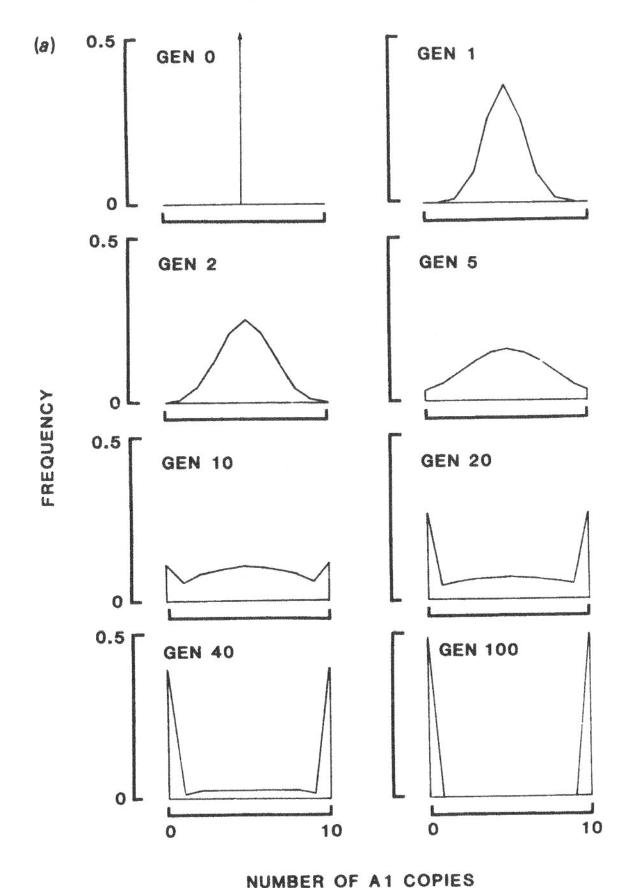

(*b*) The bivariate frequency distribution of the number of copies of an allele per nucleus and the number of copies of the locus per nucleus, when only the total number of elements in the nucleus is conserved. Each table is the triangular matrix of $p(i, j)$, the frequency of nuclei having i A1 elements out of a total of j A-elements, out of a fixed total number of elements; columns are i and rows j. The original nucleus in heterozygous, bearing a single A1 element out of a total of 2 A-elements, the total number of all elements being 4. Note the rapid dispersion of the nuclei towards the three absorbing states at $p(0, 0)$, $p(0, 4)$ and $p(4, 4)$, with the eventual fixation of these three states in the proportions 2:1:1. GEN denotes generation.

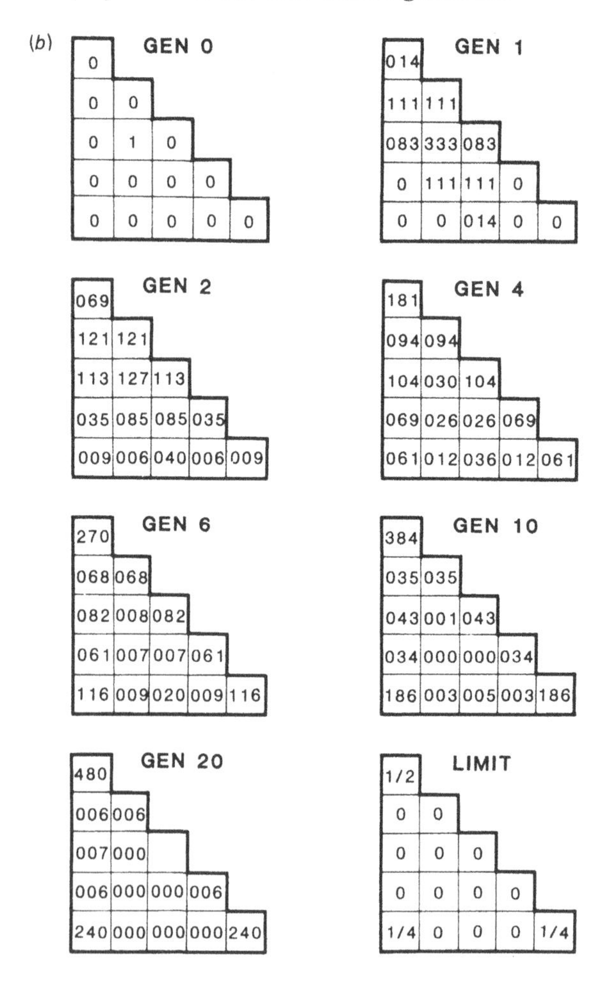
(*b*)

GEN 0				
0				
0	0			
0	1	0		
0	0	0	0	
0	0	0	0	0

GEN 1				
014				
111	111			
083	333	083		
0	111	111	0	
0	0	014	0	0

GEN 2				
069				
121	121			
113	127	113		
035	085	085	035	
009	006	040	006	009

GEN 4				
181				
094	094			
104	030	104		
069	026	026	069	
061	012	036	012	061

GEN 6				
270				
068	068			
082	008	082		
061	007	007	061	
116	009	020	009	116

GEN 10				
384				
035	035			
043	001	043		
034	000	000	034	
186	003	005	003	186

GEN 20				
480				
006	006			
007	000			
006	000	000	006	
240	000	000	000	240

LIMIT				
1/2				
0	0			
0	0	0		
0	0	0	0	
1/4	0	0	0	1/4

mechanism for regulating copy number this is unlikely to be true, and in a more realistic treatment the total number of copies M is itself a random variable. Our model then posits a nucleus with a total number of copies of all elements combined E, of which M are copies of A-elements. After replication there are $2E$ elements, of which E are chosen at random to stock one of the two new daughter macronuclei. We now have to calculate both the probability that a given daughter nucleus will receive x A-elements and the probability that y of them will be A1 elements. There are now three absorbing states: nuclei with no A1 elements, nuclei with no A2 elements, and nuclei which have lost both A1 and A2 elements. Figure 2(*b*) illustrates, for a simple case, how the frequency distribution of copy number spreads out until all nuclei are located at one or another of these three possible end-points. As in the previous case, the rate of loss of heterozygosity eventually approaches an asymptotic value, but this value is always lower than when the total number of A-elements is conserved; the presence of a sea of other types of elements slows down the rate at which a heterozygous stock assorts into homozygous lines at any given locus. This rate will depend not only on M, but also on the total number of elements E and on the proportion that A-elements comprise of the whole, M/E.

This model is in turn unrealistic; it predicts the assortment of large numbers of genetically crippled lines which lack essential elements, and this does not seem to have been observed. The truth presumably lies between the two extremes, with copies of the same element behaving neither purely dependently (as in the first model, with copy number conserved in each daughter) nor purely independently (as in the second model, with copy number a random variable). The observed rate of assortment will then estimate, not the number of copies of a given element alone, but some complex function of this variable together with the total number of elements per nucleus and the correlation between elements during amitotic division. The conventional estimate got by assuming that copy number is conserved is an underestimate of the number of copies present.

The assortment of haploid elements smaller than whole chromosomes accounts for the major features of macronuclear behaviour, and in particular for the assortment of heterozygotes into homozygous lines of nearly equal frequency at rates which are similar for different loci, whether or not the loci are being expressed, for the lack of association in assorted lines between loci known to be linked in micronuclear inheritance, and for the lack of reversion from homozygosity to heterozygosity. It breaks down, however, in two important respects. In the first place, assortment does not always begin immediately in the exconjugant lines; rather, for each locus a characteristic number of fissions must pass before any assortment can be detected. Secondly, the two homozygous lines do not always assort in equal numbers; at the *H* locus in particular, 'output ratios' are often strongly skewed in a manner which reflects phenotypic dominance, the more dominant alleles forming the majority type. These phenomena both suggest that the elements which assort are incompletely haploid. According to Nanney (1980) they are in fact diploid, one allele being irreversibly suppressed, at some characteristic number of divisions after conjugation, in each copy. Allen and Gibson (1972) prefer the idea that in each daughter macronucleus a large number of haploid 'slave' elements are proliferated by a single diploid 'master' copy. The masters are normally conserved, one entering each daughter macronucleus during amitosis, while the slaves are passively diluted out; assortment occurs as the consequence of an occasional error resulting in the formation of a macronucleus lacking a master. I do not find either of these hypotheses very compelling, but my main concern is not so much with the mechanism of somatic assortment as with its undoubted capacity to create stable and heritable phenotypic variation within a pure line and thereby to enable selection to occur. The relationship of somatic assortment to senescence is the subject of section 8/1.

4

The fate of isolate cultures

The efforts to verify Maupas' result and test Weismann's conjecture occupied the first thirty years of the century, before petering out towards the end of the 1930s. I have listed the most important of these investigations in Table 1. They represent the product of enormous labour. Must of this was ill-directed; some of the most extensive investigations, such as those Galadjieff and Metalnikov, are only sketchily described; and of course the raw data for most of the experiments is now irretrievably lost. What remains, however, is still an impressive number of long-term records in which the average fission rates of isolate lines have been reported. I have identified 75 cases in which a fresh and uniform quantitative analysis is feasible, and these are described more fully in Table 2. The million or so daily isolations analysed in this table represent the great bulk of what we know about clonal longevity from direct observation.

Unfortunately, it is difficult to analyse the fission-rate date straightforwardly. Indeed, the proposition that a clone of protozoans can be propagated indefinitely, given the right conditions, might almost serve as the model of a badly-formulated hypothesis, since it could not be affirmed in a finite experiment, and if falsified might merely demonstrate the failure of the experimenter to maintain constant favourable conditions. The survival or extinction of a line is therefore of little direct interest; what is crucial is the trend in fission rate, since a universal decline in fission rate through the history of isolate cultures would provide a conclusive demonstration of senescence. However, even the hypothesis that fission rates decline through time is not easily tested, since the pattern of decline is not specified. We have no good reason to suppose that fission rates will decline linearly, or even monotonically; conceivably, they might increase for some time after the initial isolation and then fall steeply, in which case

Table 1. *A digest of the major long-term studies of fission rate in isolate culture*

The features of the experiments listed here are as follows:

GEN maximum number of generations observed in any line.
MFR maximum mean fission rate per day recorded in any line.
CEN census interval: 1 daily; 1+ usually daily.
LIN number of lines attempted.
MED culture medium: H hay infusion, bacterized by air or transfer; HV hay infusion, bacterized by air, with change of medium during periods of low fission rate; C cereal infusion, bacterized by air or transfer; B beef extract, bacterized by transfer; S synthetic medium; P sterilized pondwater; E 'locke-egg medium'; M manure extract; C carnivore, fed on other ciliates.
TEM temperature: C controlled, O not controlled, CO not controlled but reported as nearly constant, R room temperature, RC roughly controlled near room temperature.
AMP amphimixis (in ciliates, conjugation): 0 prevented; + not prevented.
AUT automixis: 0 prevented, or known not to occur; + not prevented.
END fate of lines: D died; S stopped before extinction; E encysted.

Organism	Authority	GEN	MFR	CEN	LIN	MED	TEM	AMP	AUT	END
Ciliata: Holotricha										
Paramecium aurelia	Woodruff 1908, 09, 11, 26	5071	3.0	1+	4	HV	0	0	+	S*
	Woodruff and Baitsell 1911*a*	177	2.7	1	8	B	R	0	+	S*
	Richards and Dawson 1927	480	1.4	1	—	—	—	0	—	S
Paramecium caudatum	Calkins 1902, 03, 04	742	1.7	1+	12	HV	R	0	+	D*
	Metalnikov 1919, 22, 24; and Galadjieff and Metalnikov 1935	8704	c2	1	10	H	CO	0	+	S*
Paramecium calkinsi	Spencer 1924	420	1.95	1	16	H	C	0	0	D
Spathidium spathula	Moody 1912	218	3.1	1	3	H	0	0	+?	D*
	Woodruff and Moore 1924	1080	—	—	—	—	—	0	0	D
	Woodruff and Spencer 1924	546	4.2	1	468	BC	0	0	0	S*
Didinium nasutum	Calkins 1915	131	2.6	1	5	C	0	0	0	E*
	Mast 1917	1035	5.7	—	—	C	—	—	—	S
	Beers 1926	786	4.5	1	24	C	0	+	0	S
	Beers 1929	1384	4.9	1	4	C	CO	+	+	S

Tillina magna	Gregory 1909	548	2.5	?1	12	M	0	0	?+	D*
Glaucoma scintillans	Enriques 1905, 1916	2701	12.0	—	—	H	—	—	—	S
Colpidium colpoda	Vieweger 1918	650	3.9	—	—	—	CO	—	—	S
Ciliata: Hypotricha										
Stylonichia pustulata	Maupas 1888	316	c3	1	1	?C	R	+	+	D
	Popoff 1907									
	Baitsell 1912	572	3.1	1	12	BH	0	+	+?	D*
Pleurotricha lanceolata	Joukowsky 1898	458	c3	1	—	—	RC	+	+	S
	Woodruff 1905	448	1.9	1	8	HV	R	0	+?	S*
	Baitsell 1914	943	4.2	1	8	H	0	0	+?	D*
Oxytricha fallax	Woodruff 1905	860	3.6	1	8	HV	R	0	+	D*
	Baitsell 1914	150	2.6	1	12	B	0	0	+?	D*
Oxytricha hymenostoma	Dawson 1919	289	4.6	1	4	HV	0	0	0	D*
Uroleptus mobilis	Calkins 1919, 20, 25	349	2.5	1	80	HC	0	0	0	D*
	Austin 1927	447	2.3	1	48	PEB	0	0	+?	D*
Gastrostyla steinii	Woodruff 1905	288	2.7	1	4	HV	R	0	+?	D*
Histrio complanatus	Dawson 1926	590	1.8	?1	12	HV	0	0	0	D*
Ciliata: Heterotricha										
Blepharisma undulans	Richards and Dawson 1927	500	1.0	?1	—	—	—	—	0	S*
	Woodruff 1927, 28									
Heliozoa										
Actinophrys sol	Belar 1924	1244	3	1	—	—	—	0	0	S*
Chlorophyceae										
Eudorina elegans	Hartmann 1921	228[a]	0.25[a]	1+	—	—	—	0	0	S*
Euglenophyceae										
Entosiphon sulcatum	Lackey 1929	947	1.8	1	48	SC	OC	0	0	S*
Peranema trichophorum	Lackey 1929	171	1	1	56	SC	OC	0	0	S*

An asterisk after the last entry indicates that suitable lines are analysed further in Table 2. Note: (*a*) 228 colony generations in the B line; since the colonies of 32 cells reproduce by autocolony formation, this is equivalent to $5 \times 228 = 1140$ cell generations. Likewise, the figure for MFR represents a minimum colony generation time of 4 days.

Table 2. *The change of vitality through time in long-term isolate cultures of protists*

I have analysed all the cases for which I was able to abstract a complete time series from the literature. The dependent variable is divisions per day averaged (by the original authors) over 5- or 10-day periods; except that I have averaged Hartmann's extensive data on *Eudorina* at 50-day periods, and the *Paramecium caudatum* study conducted by Galadjieff and Metalnikov is reported as yearly averages. The independent variable is time in days from the beginning of the observations, expressed as the midpoint of the time interval used. The first three columns of the table give a reference number, species and strain. Authorities are as follows. 1 Woodruff and Baitsell 1911a; 2 Woodruff 1911; 3 Richards and Dawson 1927; 4 Calkins 1903; 5 Galadjieff and Metalnikov 1935; 6 Woodruff and Spencer 1924; 7 Moody 1912; 8 Calkins 1915; 9 Gregory 1909; 10 Baitsell 1912; 11 Popoff, in Gregory 1909; 12 Woodruff 1905; 13 Baitsell 1914; 14 Woodruff 1905; 15 Baitsell 1914; 16 Dawson 1919; 17 Calkins 1919; 18 Calkins 1925; 19 Austin 1927; 20 Woodruff 1905; 21 Dawson 1926; 22 Richards and Dawson 1927; 23 Belar 1924; 24 Hartmann 1921; 25 and 26 Lackey 1929. The statistics calculated for each case are:

RS is the Spearman rank correlation coefficient.

Q1, Q2 are the mean rate of division during the first (Q1) and last (Q2) quartiles of the period of observation.

B0, B1, B2, B3 are the coefficients of the cubic regression of division rate on time in days. The intercept B0 is given as divisions per day. The other three coefficients are given in exponential form; thus +29−3 should be read as $+(29 \times 10^{-3})$. Regressions were fitted by ordinary least-squares techniques.

Rsq is the coefficient of determination (squared parametric multiple correlation coefficient), expressing the proportion of the variance in division rate accounted for by the cubic regression model.

$F' < 0$ represents the time during which the first derivative of division rate with respect to time, F', is negative, according to the trend described by the cubic regression model. Only times up to the last time interval in the data are considered. Thus $t > 0$ indicates a negative trend throughout the course of the observations; $t > 145$ would indicate a negative trend from day 145 until the final time period of the data; $t < 145$ would indicate an increasing trend after day 145. Note that $t > 145$ does not preclude the possibility that the division rate first fell, then rose, and finally (from day 145 onwards) again fell.

TMAX is the largest value of time in the data series.

In all data columns except B0 an asterisk indicates a probability $P < 0.01$ that the estimate differs significantly from zero. (An asterisk between Q1 and Q2 indicates that they differ significantly at $P < 0.01$, using the t-test for samples with unequal variances). All values of B0 are very highly significantly different from zero.

Organism	Culture	RS	Q1	Q2	B0	B1	B2	B3	Rsq	$F' < 0$	TMAX
Ciliata: Holotricha											
1. *Paramecium aurelia*	I	−0.31	1.60	1.22	1.09	+29−3	−31−5	+84−8	0.23	$t < 182$	215
	IB	+0.25	1.09	1.17	1.12	−11−3	+16−5	−51−8	0.16	$t > 160$	215

2. *Paramecium aurelia*		+0.31*	1.25*	1.83	1.18	+12−4	−25−8	−33−11	0.13*	$t > 1245$	1245
3. *Paramecium aurelia*		−0.34*	0.84	0.71	1.07	−28−4	+59−7	−39−10	0.24*	$t < 405$	820
4. *Paramecium caudatum*	A	−0.42*	0.98*	0.43	1.07	−22−4	+13−6	−19−9*	0.51*	$t > 357$	680
	C	−0.95*	2.07*	0.37	2.31	−77−4	−19−7	−18−9	0.89*	$t > 0$	340
5. *Paramecium caudatum*		+0.30*			1.01	+62−6	−44−9	+97−13	0.09		4090
6. *Spathidium spathula*	Aa	−0.95*	3.86*	1.31	4.02	−48−4	−40−5	+20−7	0.91*	$t > 0$	135
	Ab	−0.95*	3.69*	0.88	4.04	−28−3	−64−6	+54−8	0.91*	$t > 0$	125
	Ac	−0.95*	3.55*	0.4[illegible]	3.81	−14−3	−43−5	+26−7	0.94*	$t < 125$	140
	Ad	−0.92*	3.47*	1.19	3.52	−75−4	−13−5	−66−8	0.87*	$t > 0$	105
	Db	−0.68*	2.87*	1.39	3.91	−11−2*	+19−4*	−11−6	0.75*	$t > 79$	125
	Eb	+0.23	1.79	2.00	1.81	−12−3	+26−5	−12−7	0.10	$t > 114$	140
	Fb	−0.82*	3.10	1.36	3.84	−11−2	+17−4	−13−6	0.78*	$t > 0$	80
	Fc	−0.95	2.89*	1.45	3.58	−53−3*	+58−5	−24−7	0.85*	$t > 0$	125
	Gb	+0.03	2.65*	2.37	2.37	+39−3	−14−4	+12−6	0.14	$t < 58$	80
	Kb	−0.30	1.63	1.37	1.91	−25−3	+45−5	−21−7	0.41*	$t > 99$	150
	Kd	+0.42	2.90*	3.40	3.39	−40−3	+56−5	−20−7	0.49	$t > 140$	155
	mean	−0.94*	2.87*	1.54	3.21	−23−3*	−73−6	+39−9	0.92*	$t > 0$	135
7. *Spathidium spathula*		−0.77*			1.17	+80−3	−15−4	+63−7	0.77*	$t > 35$	130
8. *Didinium nasutum*	X1	−0.21	1.94	1.57	1.82	−21−4	+60−5	−16−6	0.21	$t > 24$	55
	X2	−0.73*	2.01	1.15	1.97	+73−5	+74−6	−60−7	0.60	$t > 12$	65
	Y1	−0.71*	2.13	0.60	1.62	+63−3	−25−4	+18−6	0.59	$t > 15$	60
	Y2	−0.60	1.90	0.93	1.90	+18−3	+61−5	−58−6	0.80	$t > 14$	40
9. *Tillina magna*	A	−0.71	2.10	1.51	2.12	−16−3	+90−5	−20−6	0.53	$t > 0$	100
	B	−0.78*	1.86	0.63	2.20	−18−3	+85−6	−19−8	0.65*	$t > 0$	380
Ciliata: Hypotricha											
10. *Stylonichia pustulata*	Sb	−0.31	2.57	2.35	3.15	−44−3	+71−5	−33−7	0.32	$t > 99$	150
	Sbh	−0.63*	1.64	0.98	1.53	+13−3	−13−5	−27−8	0.50*	$t > 44$	270
	Sbhb	−0.86*	1.47	0.68	1.77	−14−3	+99−6	−55−8	0.77*	$t > 0$	130
11. *Stylonichia pustulata*		−0.07	1.31	1.31	1.73	−49−3	+10−4	−62−7	0.13	$t > 78$	180
12. *Pleurotricha lanceolata*	A	−0.75*	1.22	0.59	1.43	−50−4	−26−6	+14−8	0.63*	$t < 185$	240
	B	−0.32	1.03	0.87	1.19	−39−4	+15−6	+17−9	0.11	$t < 111$	480
13. *Pleurotricha lanceolata*	Pb	−0.87*	2.71	1.00	3.01	−65−4	+14−6	−26−9	0.79*	$t > 0$	470
	Ph	−0.81	3.33	1.12	2.58	+43−3*	−41−5*	+81−8*	0.92*	$t > 0$	250

Table 2. (*cont.*)

Organism	Culture	RS	Q1	Q2	B0	B1	B2	B3	Rsq	F′ < 0	TMAX
14. *Oxytricha fallax*	A	−0.24	1.27*	0.59	1.39	−71−4	+46−6	−62−9*	0.38*	$t > 393$	630
	B	−0.60*	1.47*	0.71	1.44	−50−6	−35−8	−21−9	0.46*	$t > 0$	350
15. *Oxytricha fallax*	Ob	−0.80*	1.76*	0.52	1.87	+31−4	−60−5	+43−7	0.71*	$t < 90$	110
	Ove	−0.73*	2.29*	0.94	2.53	−43−3	+12−4	−14−6	0.67*	$t > 0$	75
	Oh	−0.89*	1.99*	0.74	2.43	−32−3	+26−5	−14−7	0.79*	$t > 0$	100
16. *Oxytricha hymenostoma*		−0.83	3.62*	1.33	3.89	−32−3	−12−6	+66−8	0.70*	$t > 0$	130
17. *Uroleptus mobilis*	B	−0.95*	1.74*	0.04	2.01	−63−4	−28−6	+79−9	0.91*	$t > 0$	260
	C	−0.88*	1.67*	0.19	1.87	−47−4	+23−7	−28−9	0.84*	$t > 0$	310
	D	−0.83*	1.72*	0.05	1.58	+90−4	−13−5	+24−8	0.82*	$t > 40$	240
	F	−0.72*	1.70*	0.60	1.69	−65−4	+60−6	−22−8	0.64*	$t > 0$	270
	G	−0.68*	1.67*	0.27	2.20	−27−3*	+25−5*	−76−8*	0.71*	$t > 135$	250
	H	−0.67*	1.41*	0.12	1.39	+42−4	−12−6	−16−8	0.71*	$t > 72$	240
	I	−0.53*	1.60*	0.67	2.05	−24−3	+30−5*	−10−7*	0.77*	$t > 131$	230
	J	−0.76*	1.80*	0.48	1.82	−65−4	+13−5	−82−8	0.86*	$t > 62$	190
18. *Uroleptus mobilis*	23	−1.00*	1.43*	0.07	1.51	+13−3	−47−6*	+89−9*	0.99*	$t > 0$	380
	dble	−0.78*	1.13*	0.20	1.04	+52−4*	−45−6*	+66−9*	0.97*	$t > 68$	390
19. *Uroleptus mobilis*	1	−0.95*	1.10*	0.29	1.17	−15−4	+53−7	−29−9	0.74*	$t > 0$	360
	2	−0.96*	1.49*	0.20	1.51	+13−5	−39−6	+73−9	0.94*	$t > 0$	360
	3	−0.91*	1.01*	0.34	1.05	−17−4	+94−7	−32−9	0.88*	$t > 0$	380
	4	−0.89*	0.90*	0.23	1.05	−57−4*	+30−6*	−66−9*	0.90*	$t > 0$	350
	5	−0.97*	1.16*	0.12	1.27	−28−4	+40−7	−38−9	0.95*	$t > 0$	290
	6	−0.88*	1.00*	0.33	1.16	−71−4*	+48−6*	−11−8*	0.90*	$t > 0$	330
	7	−0.88*	1.11*	0.14	1.86	−33−3*	+23−5*	−50−8*	0.99*	$t > 177$	250
	8	−0.91*	1.52*	0.34	2.08	−18−3*	+80−6*	−14−8*	0.92*	$t > 0$	320
	9	−0.89*	1.01*	0.19	1.51	−14−3*	+53−6*	−70−9*	0.98*	$t > 0$	380
	10	−0.78*	0.94*	0.16	1.05	−19−5	−64−7	−45−10	0.63*	$t > 0$	370
	11	−0.61	1.24	0.31	1.47	−26−3*	+55−5*	−33−7*	0.94*	$t > 78$	140

	12	−0.93*	1.77*	0.19	1.52	+97−4	−11−5*	+21−8*	0.89*	$t < 304$	360
	13	−0.91*	2.07*	0.10	1.51	+41−3*	−66−5*	+22−7*	1.00*	$t < 166$	190
	mean	−0.99*	1.23*	0.17	1.48	−66−4*	+14−6*	−18−9	0.98*	$t < 0$	380
20. *Gastrostyla steinii*		−0.16	1.39	1.25	0.60	+53−3*	−63−5*	+21−7*	0.38*	$t < 147$	190
21. *Histrio complanatus*	A	−0.82*	1.32*	1.45	1.20	+35−4	−19−6*	+18−9*	0.70*	$t < 595$	670
	B	−0.33	0.55*	0.28	0.52	−70−5	+10−6	−40−9	0.22	$t > 117$	310
Ciliata: Heterotricha											
22. *Blepharisma undulans*		−0.35*	0.61*	0.43	0.60	+46−6	−60−8	+35−11	0.15		1000
Heliozoa											
23. *Actinophrys sol*		+0.12	1.39	0.38	1.55	−27−4*	+51−7	−26−10	0.07*	$t > 704$	990
Algae: Chlorophyceae											
24. *Eudorina elegans*		+0.37	1.60	0.93	0.81	−95−5	+94−8	−15−11	0.32	$t < 585$	1650
Algae: Euglenophyceae											
25. *Entosiphon sulcatum*	2HE	−0.65*	1.60*	0.96	1.64	+16−4	−31−6	+61−9	0.47*	$t < 316$	410
	EV	+0.39*	0.26	1.53	1.47	−61−4	+36−6	−50−9	0.20*	$t > 367$	460
	EW	−0.16	1.60	1.52	1.58	+24−5	−46−7	−67−10	0.12	$t < 450$	630
26. *Peranema trichophorum*	BP	−0.58*	0.45	0.15	0.56	−92−4	+95−6	−32−8	0.43	$t > 107$	190
	BPA	−0.50	0.70	0.47	0.70	−22−4	+45−7	+19−10	0.28	$t > 0$	190
	BPAA	+0.10	0.76	0.65	0.82	−73−4	+21−5	−13−7*	0.79*	$t > 90$	150
	APA	−0.67*	0.69	0.29	0.92	−12−3*	+10−5*	−28−8*	0.72*	$t > 141$	240
	CP	−0.50*	0.67*	0.24	1.19	−16−3*	+88−6*	−15−8*	0.69*	$t > 262$	360

the obvious procedure of linear regression would give wholly misleading results. In the absence of any usable prior hypothesis, the only reasonable course of action seemed to be to compute several measures of trend and hope that they gave concordant results.

The first step was to calculate a rank correlation coefficient. This describes the trend in fission rate with time, positive or negative, without making any assumption about the pattern of change, except that, like any single measure of correlation, it assumes the trend to be monotonic and will give a misleading picture of curves which first rise and then fall or vice versa. The frequency distribution of this coefficient among the 75 cases analysed is shown in Figure 3. It shows very clearly a general tendency for the fission rate to decline with time; 65/75 of all estimates are negative, and of the 51 estimates which differ significantly from zero (at $P = 0.05$), 49 are negative. Moreover, the negative coefficients are rather large in magnitude, with 46 falling below -0.70. There is, however, a curious hump at the other end of the distribution, between $+0.20$ and $+0.40$. There are not many cases in this region, but there certainly appear to be

Figure 3. Frequency distribution of the rank correlation coefficient. Data from Table 2; each unit square represents one case. Cases are accumulated at intervals of 0.1 units.

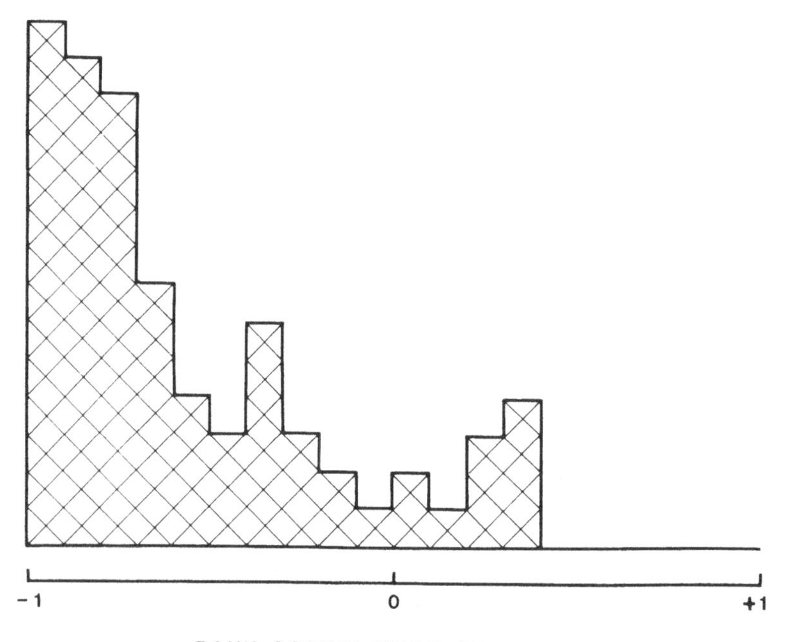

many more than we would expect from the shape of the rest of the distribution.

Another crude description of vitality can be got by dividing each data set into four equal periods and calculating the mean fission rate in the first and last of these quartiles. The result of this calculation is sketched in Figure 4. The mean fission rate is in almost all (69/75) cases lower in the last than in the first quartile, confirming a consistent tendency for fission rate to fall with time.

Since the overall pattern appeared to be clearcut, a more ambitious attempt to describe the data by regression analysis was made. Since no

Figure 4. Fission rates in the first (Q1) and last (Q2) quartiles of the period of observation. Letters indicate taxa: A *Actinophrys*, D *Didinium*, E *Eudorina*, G *Gastrostyla*, H *Histrio*, L *Pleurotricha*, N *Entosiphon*, O *Oxytricha*, P *Paramecium*, R *Peranema*, S *Spathidium*, T *Tillina*, U *Uroleptus*, Y *Stylonichia*. Data from Table 2.

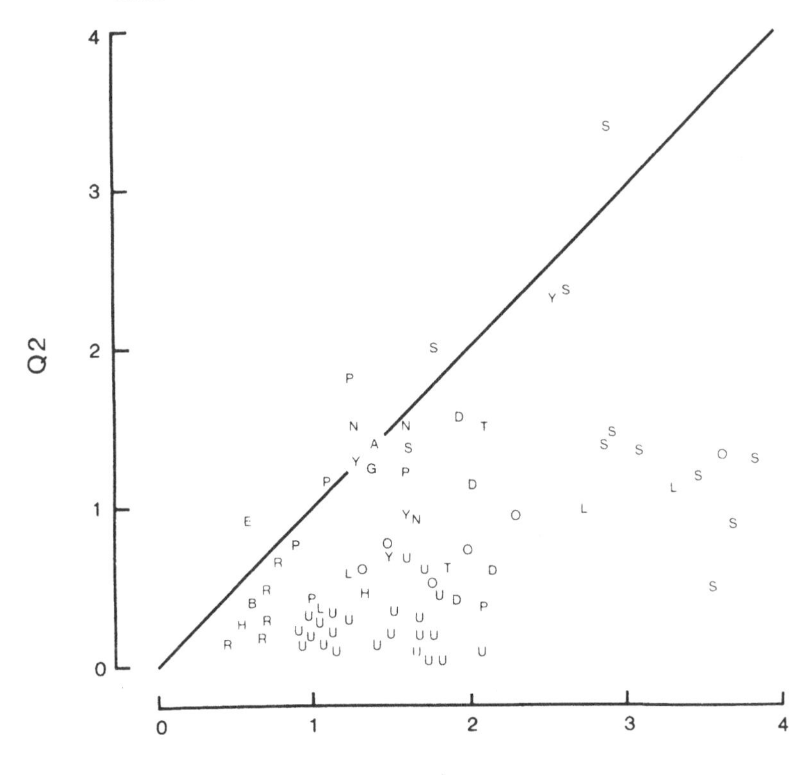

a priori model was available, I decided to relate fission rate F to time t through a cubic equation:

$$F = b_0 + b_1 t + b_2 t^2 + b_3 t^3.$$

The least-squares fit of this model is easy to compute with standard statistical packages, and in practice no generally superior model was found. Having estimated the coefficients of this model for a particular case, we can use these estimates to measure the trend in fission rate at any given time by taking the first derivative of F with respect to t:

$$\mathrm{d}F/\mathrm{d}t = b_1 + 2b_2 t + 3b_3 t^2.$$

Moreover, we can describe the way in which the trend itself changes through time by differentiating again to get:

$$\mathrm{d}^2F/\mathrm{d}t^2 = 2b_2 + 6b_3 t.$$

This procedure, then, gives us a convenient way of expressing both the trend in fission rate and the way in which this trend changes through time. Some examples of the fitted regressions – encompassing rather more than half the total data set – are shown in Figure 5.

Figure 5. Trend of fission rate with number of generations elapsed since isolation in various taxa. Each curve is the cubic regression fit to an isolate series. Numbers or symbols indicate series plotted. All the above curves are plotted out to 200 days.

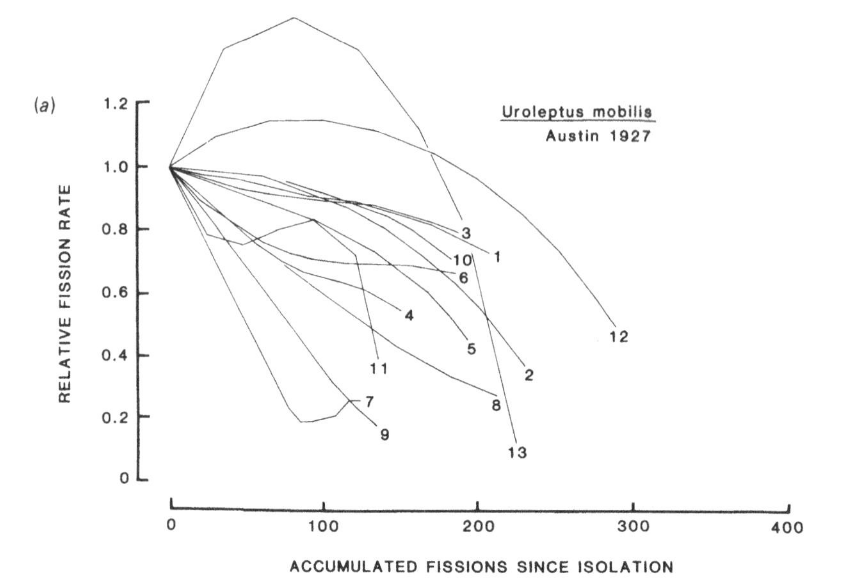

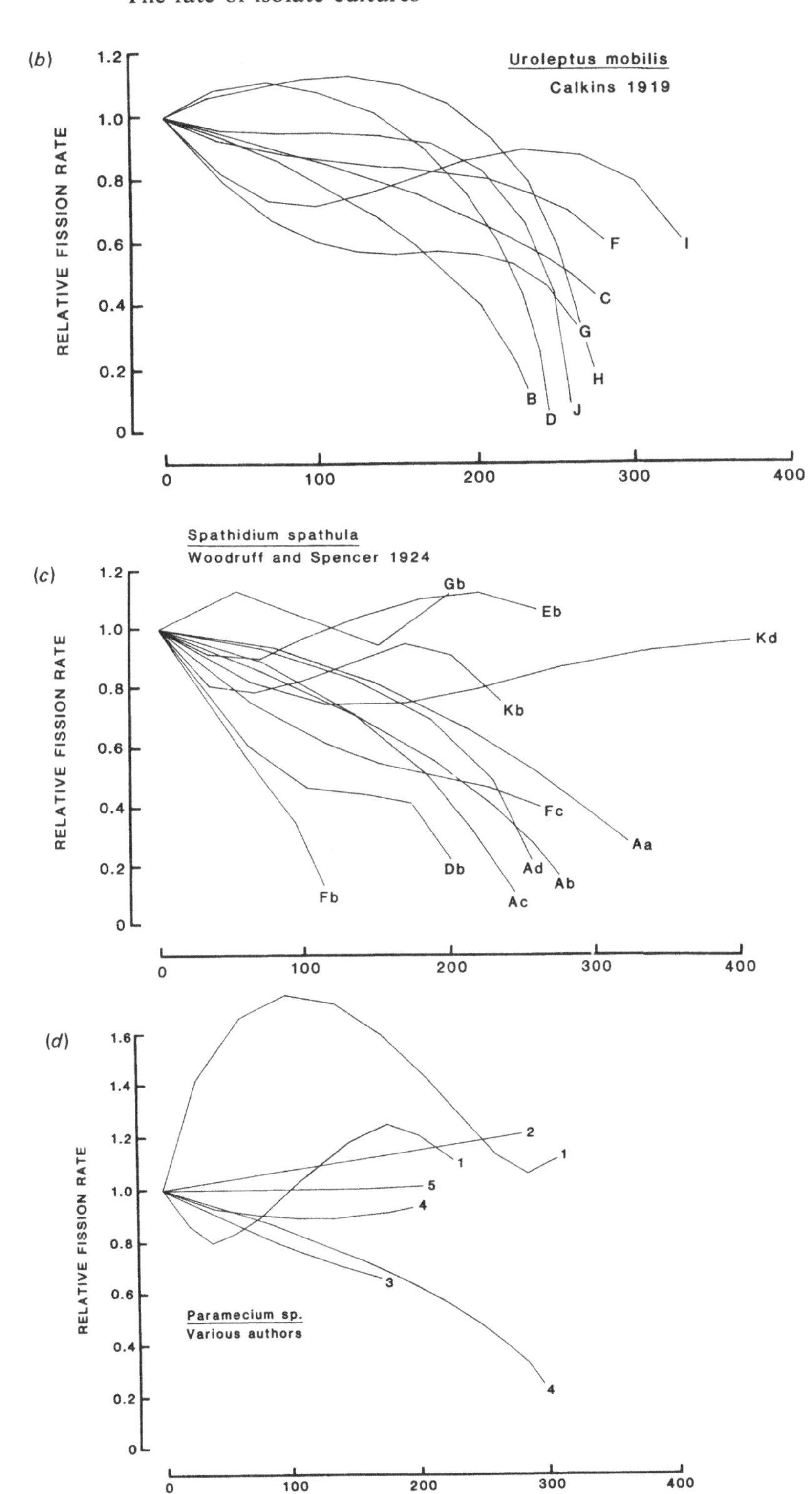
(b)
Uroleptus mobilis
Calkins 1919
RELATIVE FISSION RATE
1.2
1.0
0.8
0.6
0.4
0.2
0
F
I
C
G
H
B
J
D
0
100
200
300
400
Spathidium spathula
Woodruff and Spencer 1924
(c)
RELATIVE FISSION RATE
1.2
1.0
0.8
0.6
0.4
0.2
0
Gb
Eb
Kd
Kb
Fc
Aa
Ad
Ab
Db
Ac
Fb
0
100
200
300
400
(d)
RELATIVE FISSION RATE
1.6
1.4
1.2
1.0
0.8
0.6
0.4
0.2
0
2
1
1
5
4
3
4
Paramecium sp.
Various authors
0
100
200
300
400
ACCUMULATED FISSIONS SINCE ISOLATION

On the basis of our previous results, we expect that on average the trend will be negative, but we have as yet no information on how it changes with time. In Figure 6(*a*) the average value of the trend in fission rate, taken across all available cultures, is plotted against time for the first 200 days after isolation. Although there is obviously a good deal of variation between experiments, there is no doubt that the trend is negative at all

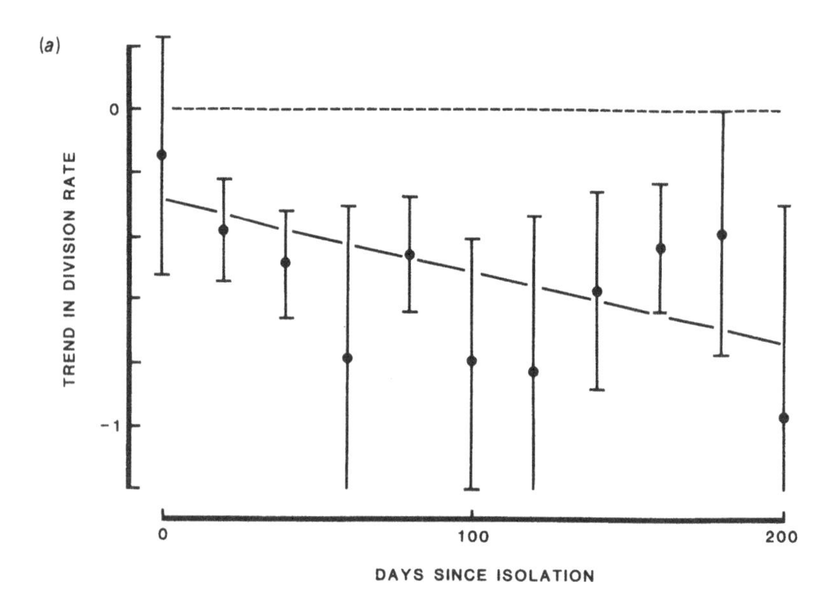

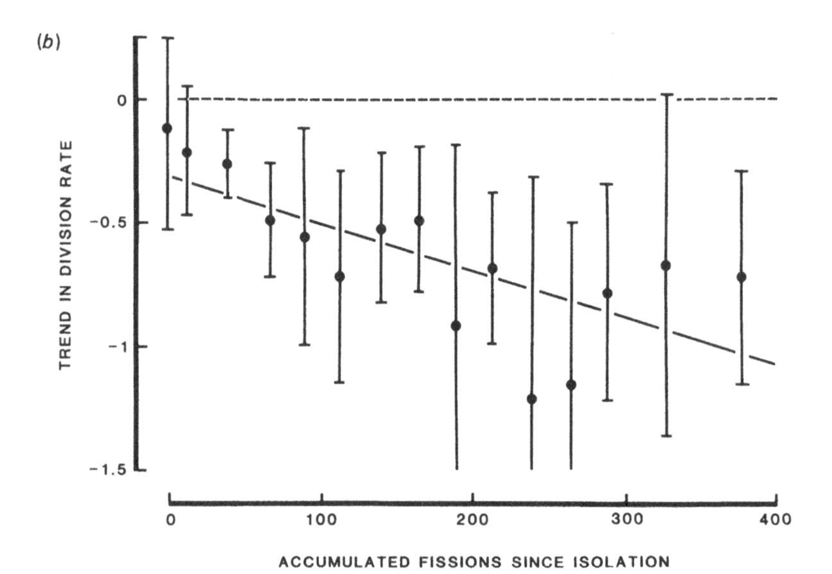

times from about 20 days after the initial isolation. Moreover, there is a strong suggestion, falling short of formal significance, that the trend to become more negative with time, indicating an accelerating decline in vitality.

It might be objected that in this analysis we are pooling organisms which operate on rather different time scales. *Spathidium*, for example initially divides three or four times daily, but *Paramecium* only once, and others still more slowly. By integrating the regression equation, we can estimate the accumulated number of fissions corresponding to a particular time in any given case. When we make this more realistic plot of fission rate on accumulated fissions (Figure 6(*b*)), the result is even more decisive: the trend, as before, is always negative, but the tendency for this trend to become more negative when more generations have passed is now highly significant. This analysis leads us, therefore, to two primary generalizations: first, that the fission rate generally declines through time, and secondly, that this decline becomes steeper in later generations. It follows that, as a general rule, isolate cultures are mortal and must eventually die out.

At the same time, the variation in the data warns us that this general rule may have exceptions, and these exceptions, if they exist, are of great importance, since even a single genuine instance of immortality must be of great interest. Besides techniques to describe general trends, therefore, we

Figure 6. Overall trends in fission rate through time. (*a*) Trend in fission rate as a function of time since isolation. Trend is calculated as $100 \times (1/F)(dF/dt)$, where F is got as a function of t using the cubic regression fit. Each point is the mean value of trend at a given time t for all those cultures surviving to time t, except that a very few cases in which negative fission rates were predicted near the end of the lifetime of a culture were excluded. Bars are 95% confidence limits of the mean. The slope of the fitted linear regression is -0.00176 ± 0.00110, for which Student's $t = 1.60$, $P \approx 0.15$; the intercept is -0.385. (*b*) Trend in fission rate as a function of the number of generations elapsing since isolation. Trend calculated as in the previous figure; for any value of t, say T, the number of generations completed was calculated as $G = \int_{t=0}^{t=T} F(t)dt$ from the coefficients of the fitted cubic regression. Each point is the mean value of trend for all values of G falling within 25-generation intervals; except that the initial value (at $G = 0$) is plotted separately, and the final two points are at 50-generation intervals in order to maintain a reasonable sample size. The slope of the fitted regression line is -0.00192 ± 0.00053, for which Student's t = 3.62, $P < 0.01$; the intercept is -0.308. Note, however, that the accumulated number of fissions is an estimate, whose variance is unknown.

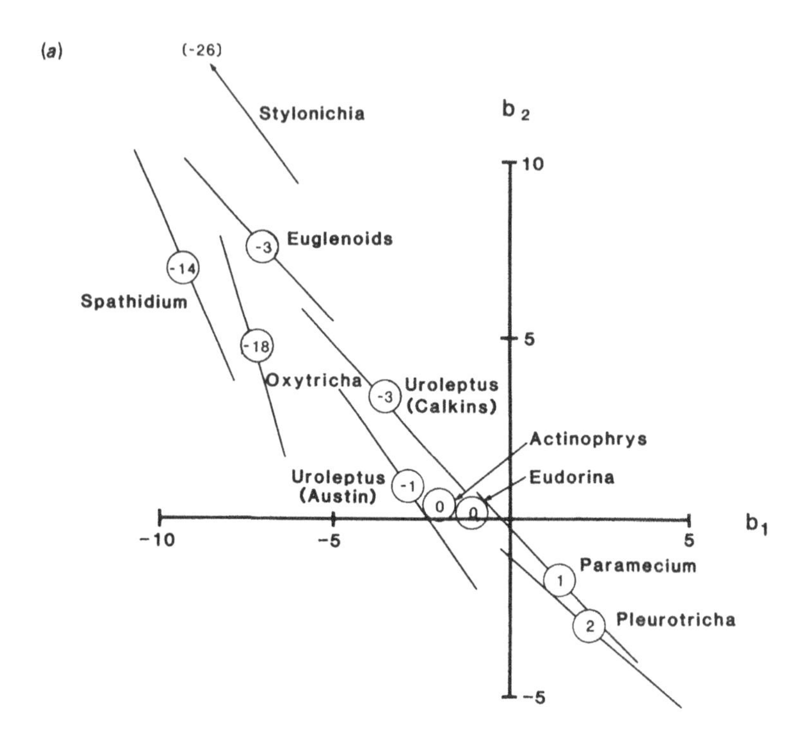
(a)
(-26)
Stylonichia
b 2
10
Euglenoids
-3
-14
Spathidium
5
-18
Oxytricha
-3
Uroleptus
(Calkins)
Actinophrys
Eudorina
Uroleptus
(Austin)
-1
0
0
−10
−5
5
b1
Paramecium
1
2
Pleurotricha
−5

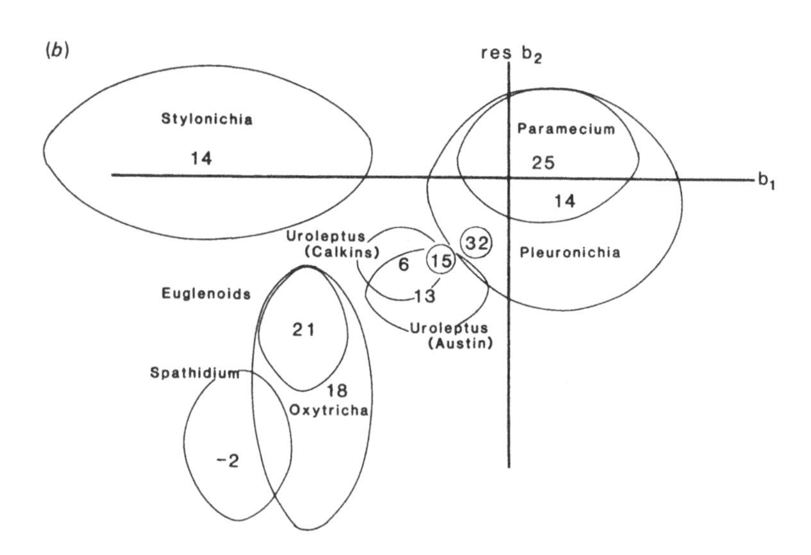
(b)
res b2
Stylonichia
14
Paramecium
25
b1
14
Uroleptus
(Calkins)
32
Pleuronichia
6
15
Euglenoids
13
Uroleptus
(Austin)
21
Spathidium
18
Oxytricha
−2

also need techniques which display the variety of behaviours shown by the cultures. I have tried to draw maps which illustrate this variety by using the results of the regression analysis.

The most straightforward procedure is to use the regression coefficients directly to describe the differences between cultures. To make different organisms more directly comparable, I have first divided the estimates of b_1 and b_2 by the intercept b_0, to obtain coefficients which are in each case relative to an initial fission rate of unity. These coefficients are plotted in Figure 7(*a*) for some of the most thoroughly investigated organisms. This makes it clear that the estimated coefficients are not a random sample of all the possibilities: in most cases there is a tendency for b_1 to be negative and b_2 positive; at the same time, the diagram also shows that b_3 is generally negative. In fact, 41 of the 75 individual cultures (55 %) show this pattern of $-/+/-$ for the coefficients b_1, b_2 and b_3, many of the remainder (16/34, or 47%) being $+/-/+$, and most of the other six possible combinations occurring infrequently or (in the case of $+/+/+$) not at all. Moreover, different organisms map in different regions of the diagram. *Spathidium* and *Oxytricha* generally have large negative values of b_1 and b_3 and large positive values of b_2; *Paramecium* and *Pleurotricha*, on the other hand, have small positive values of b_1 and b_3 and small negative values of b_2. Both major series of *Uroleptus* are intermediate between these two extremes, with moderate negative values of b_1, moderate or small positive values of b_2 and small negative values of b_3.

A very striking feature of Figure 7(*a*), and the principal weakness of using direct plots of this sort, is the high degree of correlation between estimates of regression coefficients. The first two coefficients, b_1 and b_2, are negatively correlated both between and within the series; b_3 tends to be negative when b_1 is negative and b_2 positive, but positive when b_2 is negative and b_1 positive. This pattern is largely a statistical artefact, since

Figure 7. Relationships between regression coefficients of time series. (*a*) Plot of raw regression coefficients. Each coefficient (b_1, b_2, b_3) is divided by b_0 to obtain a value relative to $b_0 = 1$; b_1 is multiplied by 10^3, b_2 by 10^5 and b_3 by 10^7 to make all three coefficients roughly equal in magnitude and distributed between about -100 and $+100$. The position of each taxon in terms of the mean of these reworked values of b_1 and b_2 is indicated by a circle in which the average value of b_3 is written. The line passing through each circle represents the regression of b_2 on b_1 for the taxon. (*b*) Plot of residual regression coefficients. These are calculated as described in the text. The bars on the axes are both at a value of -10. Format as in the preceding figure, except that the ellipses by which each taxon is represented are 50% confidence limits of the bivariate mean.

estimates of regression coefficients are autocorrelated. Suppose, for example, that we were to sample a random cloud of points representing two uncorrelated variates. Any finite sample we take from this cloud will, by the accidents of sampling, yield a non-zero regression coefficient, negative in some cases and positive in others. The bivariate mean (the intersection of the true average values of the $x-$ and $y-$ variates) will be rather well-estimated by even a small sample of points, and the regression line will therefore pass close to it. Suppose that our estimated slope is negative. Since the regression line is in any case anchored somewhere close to the true bivariate mean, the intercept on the y-axis is likely to be greater than would be the case if the true slope of zero had been found. Conversely, if we had found a positive slope, the y-intercept would be estimated as smaller than its true value. If we sampled this population repeatedly, therefore, we would discover that our estimates of intercept b_0 and slope b_1 were negatively correlated. This correlation has no biological significance; it is merely an artefact of sampling. It will appear even if the $x-$ and $y-$ variates are themselves correlated, and correlating one regression coefficient with another should therefore be attempted only with some caution.

This explains why, when we plot b_1 on b_2 for the fission rate data we find a steep and highly significant relationship. This has two sources. The first is the statistical autocorrelation I have just described. This produces a graph with negative slope that should pass through the (0, 0) origin of the plot, since there is no statistical reason for zero b_1 to be associated either with positive or with negative b_2. The second is a possible biological effect, which, over and above any effect of sampling, might cause high values of b_1 to be associated with low values of b_2 or vice versa. This effect can be detected from the elevation of the line, which will be raised above or below the origin at (0, 0). To measure this effect, I first plotted the estimate of b_1 (relative to $b_0 = 1$) on that of b_2 for each case, and fitted a linear regression. The value of each b_2 is then recalculated as its deviation, positive or negative, from this regression line. To this value is then added the value, positive or negative, of the $y-$intercept. The result is a residual value of b_2, free from at least most of the confounding effects of statistical autocorrelation and absolute differences in division rates. To calculate a similar residual value for b_3 is more complicated, since we must take into account its autocorrelation both with b_1 and with the residual value of b_2, but follows a similar recipe.

The relativised residual regression coefficients obtained in this way are now virtually uncorrelated with one another and can be used to describe the underlying differences between cultures. Comparing this treatment

(Figure 7(*b*)) with the original (Figure 7(*a*)), we can see that the plot of the raw regression coefficients has been, roughly speaking, rotated through 90° as the result of removing the negative autocorrelation in the data. The diversity of behaviour, however, has been conserved. *Uroleptus* is an appropriate reference point. There are two fairly extensive independent series of observations, by Calkins and Austin, which map close to one another. Both show negative first-order (b_1) and second-order (residual b_2) effects, indicating an underlying tendency for an accelerated decline in fission rate with time. Woodruff and Spencer's eleven *Spathidium* cultures show more extreme behaviour, with large negative first- and second-order effects indicating a precipitous decline in vitality. The smaller quantities of data available for *Oxytricha* and Lackey's euglenoids indicate a position intermediate between *Uroleptus* and *Spathidium*. *Stylonichia* shows consistently negative first-order effects but little or no second-order effect. *Pleuronicha* has in most (3/4) cases small negative first- and second-order effects, but the single exception has rather large positive first- and second-order terms, causing the genus as a whole to map near *Paramecium*, close to the (0, 0) origin of the graph. The residual value of the third-order coefficient b_3 is in almost all cases positive, though it appears to be zero or negative for *Spathidium*. It must be borne in mind that the coefficients mapped in this figure are no longer predictive – they do not, that is, describe the change in fission rate through time – but rather attempt to isolate different components, acting at different times, of the general decay of vitality. The plot shows that fission rates initially decline; that as time goes on there is a tendency for this decline to steepen; and that after a considerable period of time, when most cultures have developed very low rates of fission, the rate of decline tends to level out.

In the final diagram, I have tried to bring out a further difference between the cultures. Although each set of regression coefficients attempts to predict the pattern of change in fission rate, the success of these attempts varies widely; in some cases the data are fitted very closely by the calculated curve, while in others there is a great deal of variation around the regression line. To express the success of the analysis we can calculate in each case the proportion of the total variance in the data due to the regression, a quantity equal to the square of the correlation coefficient and known as the 'coefficient of determination'. In Figure 8 the trend of the fission rate 50 generations after isolation is plotted against the coefficient of determination. *Spathidium* again appears to represent an extreme case, having not only a highly negative trend but also generally large coefficients of determination, showing that fission rate in this organism not only declines steeply with time but also that it does so in a very regular fashion.

Williams (1980) has recently published a study of clonal aging in *Spathidium*, half a century after the classical work of Woodruff: it shows the same regular, positively accelerated decline in vitality. *Oxytricha* also combines negative trend with a high degree of determination; in *Uroleptus* the trend is slighter but the determination even stronger. On the other hand, forms such as *Paramecium*, *Eudorina* and *Actinophrys*, which

Figure 8. Scatter-plot of trend on determination. Trend is calculated as $100 \times dF/dt$ at 50 generations after isolation. Determination is the squared correlation coefficient for the fit of the cubic regression model to the data. Some taxa are indicated by broken lines enclosing symbols: ○ Spathidium, ● Uroleptus, □ Oxytricha, + euglenoids, △ Paramecium, * Actinophrys and Eudorina. Other cultures are indicated by numbers which correspond to their order of appearance in Table 2.

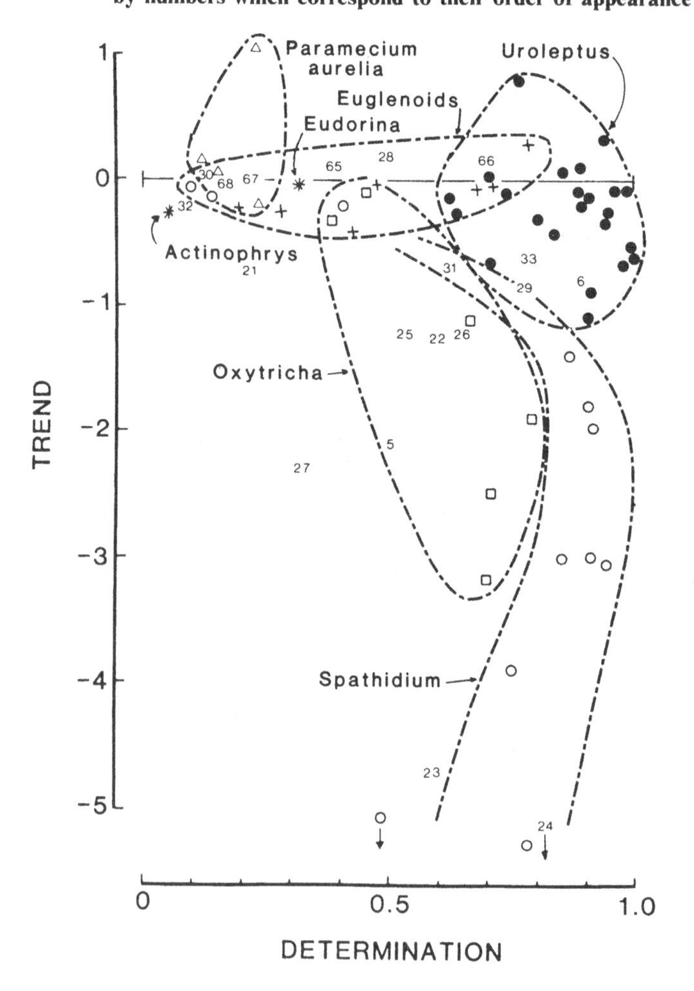

previous analyses have shown to have little trend, also have lower coefficients of determination. Indeed, there is a general tendency for trend and determination to be negatively related. The separation of forms such as *Spathidium* and *Paramecium* is thereby strengthened, fission rate declining not only steeply but also very regularly in the former, while it declines slowly, if at all, and irregularly in the latter.

Apart from Woodruff's *Paramecium*, a few cultures of ciliates show no sign of senescence. *Tetrahymena* is often said to be immortal; for example, the strain originally isolated by Lwoff in 1923 is still flourishing, despite, apparently, a complete absence of sexuality – indeed, it is now amicronucleate, and has probably been so for some time. However, like other strains for which immortality has been claimed, it has been maintained only in mass culture. The only isolate culture work with *Tetrahymena* appears to be that published by Nanney (1959), who maintained lines for about 800 generations with no consistent increase of mortality in seven of the eight clones he studied. However, the rapid growth-rate of this relatively small ciliate meant that several thousand individuals were usually produced before reisolation. A more straight-forward exception to the general rule is Beers' culture of *Didinium* (Beers 1929), which went through 1384 generations in 362 days with no sign of a decline in vitality.

Information from protists other than ciliates is regrettably scarce. The englenid *Entosiphon* passed nearly 1000 generations without senescence (Lackey 1929); however, despite the statements in Lackey's paper, *Peranema* generally declined. Although these were wide variations in fission rate during the experiments – temperature was not controlled, and some of his procedures are poorly described – there seems to be a consistent decay in fission rate when several series are plotted on the same graph (Figure 9(*a*)). Hartmann's culture of *Eudorina* was also extremely variable, especially earlier in the study, but there is no consistent decline in vitality. Since *Eudorina* is a colony of 32 cells, these observations really refer, not to an isolate culture, but rather to a small population. Belar's *Actinophrys* maintained a more or less constant rate of fission for about 1000 generations (Figure 9(*b*)).

Cultures of amoebas have produced results which are difficult to interpret. According to Muggleton and Danielli (1958; Danielli and Muggleton 1959), clones of *Amoeba proteus* and *Amoeba discoides* are 'immortal' when supplied with an excess of food; they are at any rate very long-lived, and no senile changes were observed. However, if the amoebas are maintained on a minimal diet, so that cell division does not occur, and then restored to unlimited food, the further lifespan of the clone now

Figure 9. Cultures of two non-ciliate protists. (*a*) Five series of *Peranema*, from Lackey (1929). For each series, fission rate at a given time after isolation is plotted as a proportion of the initial fission rate; the absolute values of initial fission rate were in the range 0.5–1.2. The five series are: ● BP, ○ BPA, △ BPAA, □ APA, + CP. The asterisk denotes the common origin at unity. (*b*) Belar's main culture (A-series) of *Actinophrys* (Belar 1924). The curve is the cubic regression reported in Table 2.

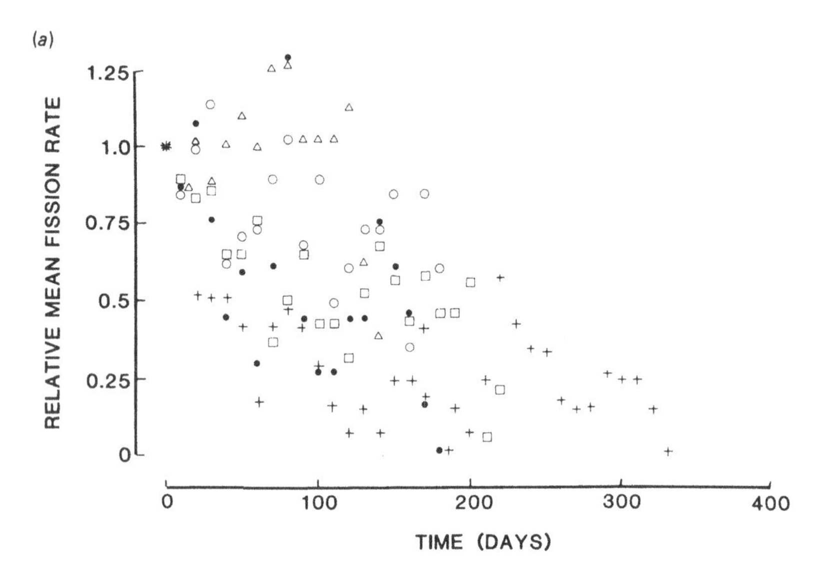

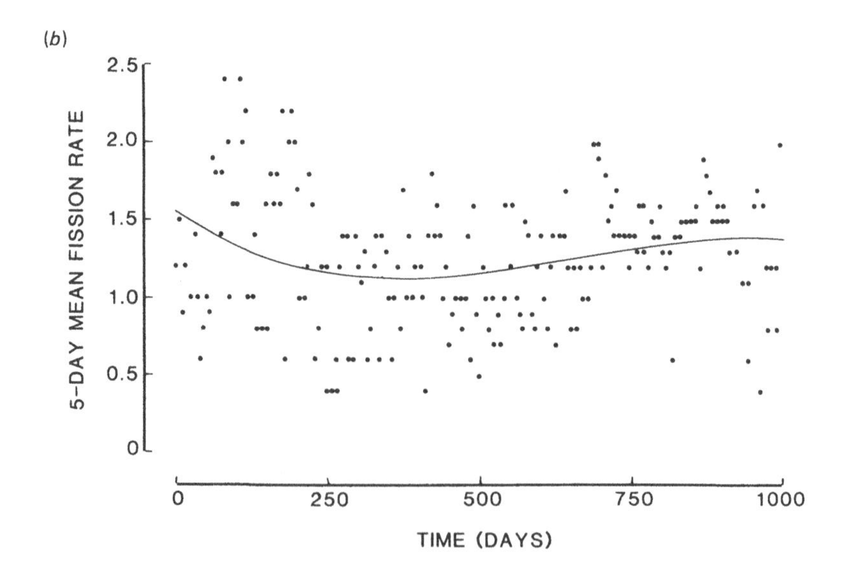

appears to be limited to 6–9 months. These 'spanned' lines are of two kinds. In the first kind, only one of the two daughter cells formed by fission is viable, the other dying soon after separation. The line eventually becomes extinct when both daughters die. In the second kind, both daughters are viable, though both multiply more slowly than individuals which have never been on a minimal diet. After some months, all the cells derived since restoration to a full diet begin to die, and all are dead a few weeks later. Muggleton and Danielli (1968) have also shown that the transfer of either nucleus or cytoplasm can transform normal into spanned amoebas, though the reverse transformation could not be accomplished. Recent results by Leith (1984) cast some doubt on earlier work. He found that about 4% of well-fed amoebas were inviable, but that a large proportion of cases involved twin sisters. Since he did not publish the genealogy of his stock, however, it is not possible to support any particular genetic hypothesis.

To sum up, the analyses I have described above establish the fundamental result that the general trend in fission rate is unarguably negative. Unless the methodology of these experiments can somehow be impugned, this fact alone seems sufficient to settle the old controversy: Weismann was wrong, and isolate lines of protozoans are not immortal. The result is the more striking since, as I have pointed above, the main source of bias in the experiments should be the unconscious selection of variants with high fission rates, which would tend to obscure any downward trend in vitality. The decline in fission rate appears to accelerate with time, or, more precisely, with the number of generations which have elapsed since isolation. After a sufficiently long period of time, however, this effect appears to decay, and the decline in fission rate slows down. Although very general, senescent decline is not universal: some cultures of *Paramecium*, *Didinium*, *Eudorina* and *Actinophrys* have been propagated for hundreds or thousands of asexual generations without any perceptible decline in fission rate.

5

The culture environment

5.1 The fate of cultures in a constant environment

Imagine a simple organism reproducing by fission; in each short period of time the probability that it will divide is b and the probability that it will die is d. If we arrange matters so that density remains constant – for instance by transferring offspring to isolate culture as soon as they are produced – then population growth will be exponential; after t units of time the expected number of individuals descending from a single founder will be $\exp(b-d)t$. However, because b and d are defined as probabilities, this value of $\exp(b-d)t$ is only the expectation of a stochastic process; the number of descendants we would actually observe in a given trial would be a random number chosen from a distribution with a mean $\exp(b-d)t$ and a variance which is also a function of b, d and t. There is even a finite probability that the number of descendants would be zero, the culture having become extinct; this is obviously so, since the original founder might die before dividing, while, if it did divide, its two decendants might both die before dividing, and so forth. In fact, the probability that a culture initially comprising n individuals will become extinct sooner or later can be shown to be:

$$\text{Prob}(0,n) = \begin{matrix} 1 & \text{if } b \leqslant d \\ (d/b)^n & \text{if } b > d \end{matrix}$$

A culture may therefore suffer stochastic extinction even when senescent changes in b and d are entirely absent. This possibility was overlooked by people working on isolate cultures, who seem to have accepted that the eventual extinction of a line was convincing evidence of senescence. We must therefore examine purely stochastic extinction as a possible explanation of the finite lifespan of many isolate lines. while remembering that it cannot account for secular trends in the fission rate.

In fact it is not straightforward to calculate the probability that a series will go extinct, because the transfer of individuals from a persistent line to restock lines which have died out means that the lines are not independent. What I have done instead is to write a computer programme that simulates the behaviour of an isolate culture. Each of four or five lines is initially stocked with a single founder. In subsequent intervals of time, each representing an hour, the programme decides whether or not an individual should survive by generating a random number between zero and one; it survives only if this number is greater than a preassigned hourly death-rate. Should it survive, the same sort of procedure decides whether or not it will reproduce, granted that it has not already undergone fission very recently. After 24 such periods, each line is reduced to a single individual, representing a daily routine of reisolation; if the line has become extinct, it is restocked from another line, if there is one with individuals to spare. The consequences of given birth- and death-rates can then be investigated by running the simulation until the whole series becomes extinct. In simple cases, we can readily confirm the results of the simulation by direct calculation. For instance, given an hourly death-rate of 0.005, suppose that the fission-rate is zero. The probability that a single line will be extinct after time t is then $P(0, 1, t) = 1-(1-d)^t$. For n lines, each line is independent (since with no fissions there can be no transfers), and the probability that all n lines will be extinct after time t is just $P(0, n, t) = [P(0, 1, t)]^n$. The median time for the extinction of all n lines is then got by solving $P(0, t, n) = 0.5$, giving $t = -(1/d) \log(1-0.5^n)$. For a series with four lines, this yields $t = 15.3$ days, while the observed mean of 20 stimulations was 15.5, with a standard error of 2.14.

The results for non-zero fission-rates are given in Figure 10, which expresses the mean time to extinction of a series as a function of the rate of fission for an hourly death-rate of 0.005. Both the mean and the variance of the time to extinction increase with the fission-rate; indeed, the variance increases faster than the mean, so that the variance/mean ratio is greater for higher rates of fission. This is because when the fission rate is very low each line is more or less independent of the other lines, since few transfers will be possible; then, since the incidence of deaths is random, the waiting-time to extinction in each line will be exponentially distributed, and the waiting-time to extinction of a series of lines will be more or less Poisson-distributed, with a variance equal to the mean. As the fission rate increases the lines become less independent, since many transfers take place to restock extinct lines, and the whole series has an exponentially distributed waiting-time to extinction, with a variance equal to the square of the mean.

More striking is the steep rise in the mean time to extinction as the fission rate increases, which suggests that stochastic extinction will take several thousand generations even for rather modest rates of increase. For instance, an hourly fission-rate of 0.025, opposed by a death-rate of 0.005, yields only $\exp(24 \times 0.02) = 1.6$ cells after one day, which would be

Figure 10. The schedule of extinction in the simulated isolate cultures. Plotted points are mean (±2 s.e.) time to extinction of 20 replicates, with a probability of death per hour of 0.005 and various fission probabilities. The regression equation is

$$\log t_E = 1.24 + 67.8F \quad (r^2 = 0.966)$$

where t_E is the mean time to extinction, in generations, and F is the probability of fission per hour. It was assumed that the probability of fission was zero in each of the 10 hours immediately following a fission.

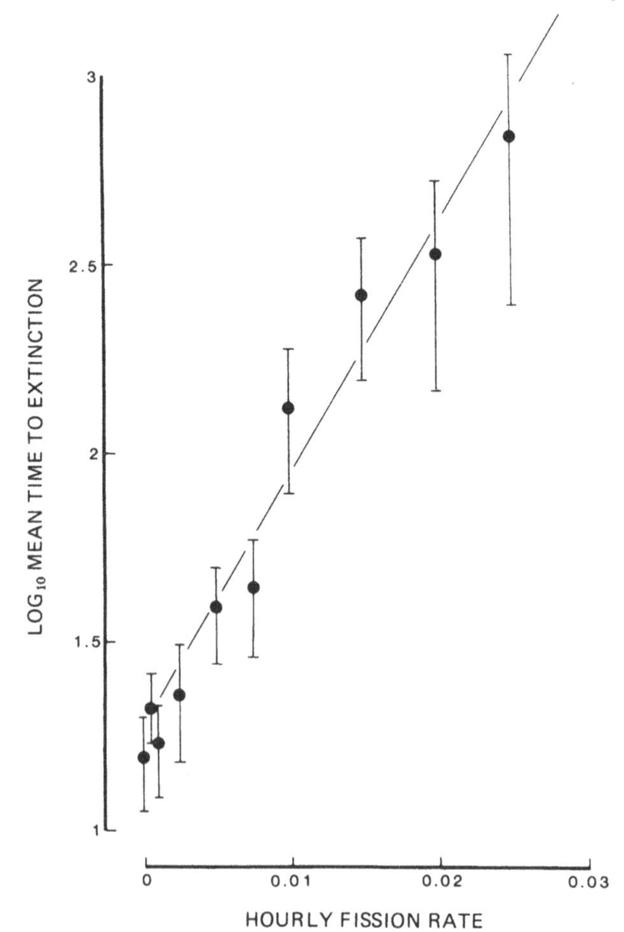

recorded as an average net fission rate of $\log_2(1.6) = 0.7$, considerably below the values measured for most ciliates; even so, the mean time to extinction will be about a thousand days, or 700 cell generations. Even this is an underestimate, since the data shown in Figure 10 include all extinctions, no matter how soon after isolation they occur; in practice, lines which became extinct a few days or weeks after isolation were usually discarded.

We can make rough estimates of birth- and death-rates for a particular case using the raw data for *Paramecium aurelia* published by Woodruff (1911) and discussed more extensively in the next section. The expected number of cells produced in one day is $\exp(24(b-d)) = 2612/995 = 2.625$ for this data, showing that $b-d = 0.0402$. The probability of death, ignoring terms of small magnitude, will be $d+(1-b)(1-d)+(1-b)^2(1-d)^2+ \ldots = d/(b+d)$ approximately, which is estimated by $53/995 = 0.0533$. Substituting one expression into the other, we obtain $b = 0.0426$ and $d = 0.0024$. Even if d were as high as 0.005, the regression shown in Figure 10 predicts that over 13000 generations would elapse, on average, before the extinction of a series. When the simulation was attempted with $d = 0.0024$, it ran out to 50000 generations without extinction occurring.

Stochastic extinction is not an attractive interpretation of the short lifespan of so many isolate cultures, given the way in which they were maintained. Even with rather low birth-rates and high death-rates, series in which individuals are transferred on occasion between four or five lines can be expected to persist for years, and extinction after a few hundred generations should not be routinely observed.

5.2 Rhythms and cycles

One of the most persistent claims in the older literature is that even in constant conditions of culture there are regular short-term cycles in the fission rate. This would imply that – contrary to the assumption made in designing the simulation – the probabilities of birth and death in successive intervals of time are not independent. The claim is not, on the face of it, very convincing, since no statistical analysis was attempted, and the ease with which structure can be read into random time series is notorious. Nevertheless, if it were true it would be very interesting, since it might provide a clue as to how some cultures avoid extinction while others do not. Rather to my surprise, it does turn out to be true.

To investigate the reality of the short-term 'rhythms' in the fission rate described by Woodruff and his collaborators, I turned to the largest source of raw data, the daily fission rates for all lines of two series of

Paramecium aurelia published by Woodruff in 1911. If we calculate the mean fission-rate per day at 5-day intervals, we have a series of 250 data-points. If rhythms do occur, these mean fission rates should exhibit a fairly regular rise and fall. We can test this hypothesis by autocorrelation, which involves calculating the correlation between division rates at all times t and those at all times $t+L$, where L is a time-lag which we can specify. A regular cycle of fission rates will be betrayed by a fall in the correlation coefficient as L increases and the cycles move out of phase, followed by a rise as L increases further and the cycles move back into phase, and so forth. This is precisely the behaviour shown by the autocorrelegram (Figure 11); the correlation coefficient reaches a peak at lags of 7–8, 19, 28, 37 and 46 time units, indicating a rather regular cycle in fission rates with a period of 9 time units, or 45 days. I can find nothing in the protocol used by Woodruff that would produce such a pattern as an artefact; rhythms appear to be a reality.

Although no other data set is suitable for autocorrelation analysis, a cruder nonparametric test suggests that such rhythms may be widespread. Suppose that we express a series of fission rates simply in terms of plus and minus signs, according to whether a given fission rate is greater or less than the preceding rate. Then the expected number of runs of like sign in a random set of N observations is $(2N-1)/3$, and the occurrence of fewer

Figure 11. The autocorrelegram for Woodruff's *Paramecium aurelia* culture. The raw data were a continuous sequence of 249 5-day means of fission rate in the series. The existence of an oscillation with a period of 9 time-units, or 45 days, is plain. Source: Woodruff (1911), Table 1.

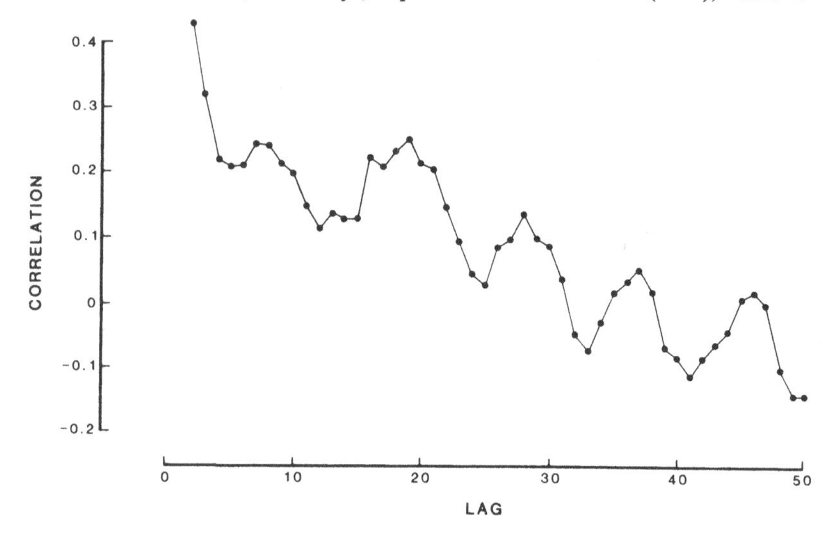

runs indicates a tendency for positive to be followed by positive and negative to be followed by negative sign, indicating the presence of a more or less regular oscillation. In 16 data sets for ciliates (excluding Woodruff's *Paramecium aurelia*), 11 showed fewer runs than expected. The total number of runs was 437 in a total of 711 observations, against the expected number of 473.7; since the variance of the expected number is known to be roughly $(16N-29)/30$, the deficiency of runs in the data is formally significant (at $P = 0.05$) if we predict, on the basis of Woodruff's results, that such a deficiency should exist.

The generation of rhythms can be investigated further by constructing the transition matrix for Woodruff's culture. This is a table which expresses the probability of passing from any given present state to any given future state; I have shown it diagrammatically in Figure 12. There is a clear lack of independence between the successive states of a given line. An individual which, on a given day, does not divide is four times more likely to die and twice as likely to survive but fail to divide on the following day as an individual which does divide. Failure to divide on a particular day is therefore not merely the outcome of a stochastic trial based on fixed probabilities of birth and death, but reveals a physiological deficiency which persists through time.

We can take this observation further. The three possible states on any particular day are d (dead), 0 (survived but did not divide) and F (divided). We can represent the successive states on days t and $t+1$ as F/d, for instance, showing that an individual which divided on day t then died on day $t+1$. What we have shown so far is that 0/0 and 0/d are much more frequent than one would expect by chance, given the overall frequencies

Figure 12. The transition probabilities between different states in Woodruff's *Paramecium aurelia* culture (Woodruff 1911, Table 1).

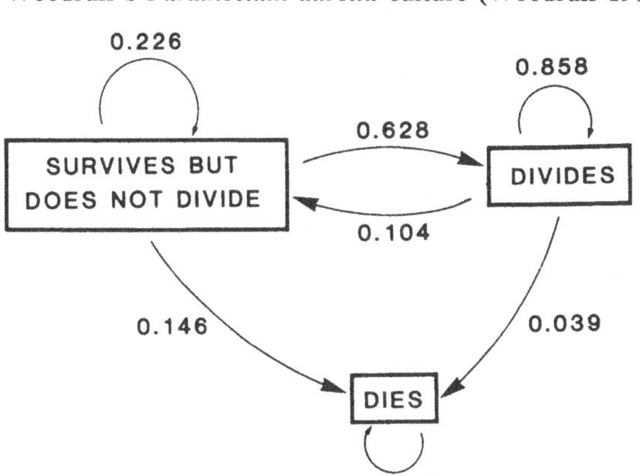

Table 3 *The effect of past on present fission rates*

Entries in columns 3–5 are transition probabilities, as functions both of the current state and of the immediately preceding state of the line. Symbols are: d = died, 0 = survived but did not divide, F = divided. MF is the mean fission rate of surviving individuals, given ±1 s.e. Columns 7–8 are X_1^2 (with Yates' correction) for the two independent comparisons of number dying versus number surviving (column 7) and division versus failure to divide among those surviving (column 8): $^{**}P < 0.01$, $^{***}P < 0.001$, ns $P > 0.05$. A comparison of the second with the fourth rows (i.e. F/0 with F/F) shows that there is a difference in the death-rate ($X_1^2 = 9.35$, $P < 0.01$), but no effect on fission among survivors ($X_1^2 = 0.01$, ns). Source of data: Woodruff (1911), Table 1.

State at:		State at $t+1$:				X_1^2 *for comparison*:		
t	$t-1$	d	0	F	MF	d versus 0, F	0 versus F	N
0	0	0.348	0.435	0.217	0.400 ±0.158			23
						8.03**	15.97***	
0	F	0.097	0.125	0.778	1.127 ±0.090			72
						2.17 ns	0.02 ns	
F	0	0.018	0.161	0.821	1.215 ±0.082			56
						0.00 ns	0.40 ns	
F	F	0.023	0.121	0.855	1.306 ±0.030			602

of d, 0 and F. If this is the result of a persistent physiological deficiency, then when we examine runs of three successive days $t-1$, t and $t+1$, we expect to find that the state at $t-1$ will influence the state at $t+1$, leading to an excess of 0/0/0 and 0/0/d observations. Table 3 shows that this is true. There is a trend in mean fission rates which suggests an inhibition of division at $t+1$ associated with previous failure to divide either at t and at $t-1$, mean fission rates at $t+1$ increasing in the sequence 0/0, F/0, 0/F and F/F, for behaviour at $t-1$ and t respectively. The major feature of the table, however, is the dramatic effect of failure to divide in both the preceding days, which substantially elevates the death rate and depresses the fission rate among surviving individuals.

5.3 The culture environment

It is certain that the attempts to maintain constant conditions of culture were often, perhaps always, unsuccessful. The evidence for this is that independent cultures maintained in the same laboratory often showed synchronous changes in fission rate. This was noticed by Calkins (1903), Woodruff (1911), Woodruff and Baitsell (1911b), Spencer (1924) and Dawson (1926). To provide an example, I have taken the data for two series of each of two different genera, *Pleurotricha* and *Oxytricha*, published by Woodruff and Baitsell (1911). For two independent comparisons between these genera, it turns out that about half the variance in the fission rate of one is accounted for by variance in the fission rate of the other (Figure 13). This must imply that some feature of their common environment – in this case, almost certainly temperature – was affecting all the cultures in the same way.

The effect of environmental variance on the fission rate will be to reduce the long-term geometric mean fission rate and thus increase the probability of extinction. Much more important, however, is the possibility that some features of the culture environment may continually deteriorate; senes-

Figure 13. The synchronous behaviour of independent cultures. Comparison of cultures of different genera of ciliates maintained in the same room. Solid circles are *Oxytricha* series *A* versus *Pleurotricha* series *A*; open circles are corresponding values for *B* series. The values plotted are deviations from the mean as proportions of the mean, for different dates. The lines are the two regressions of the pooled data; they have $r^2 = 0.50$. Source: Woodruff and Baitsell (1911b), Fig. 5.

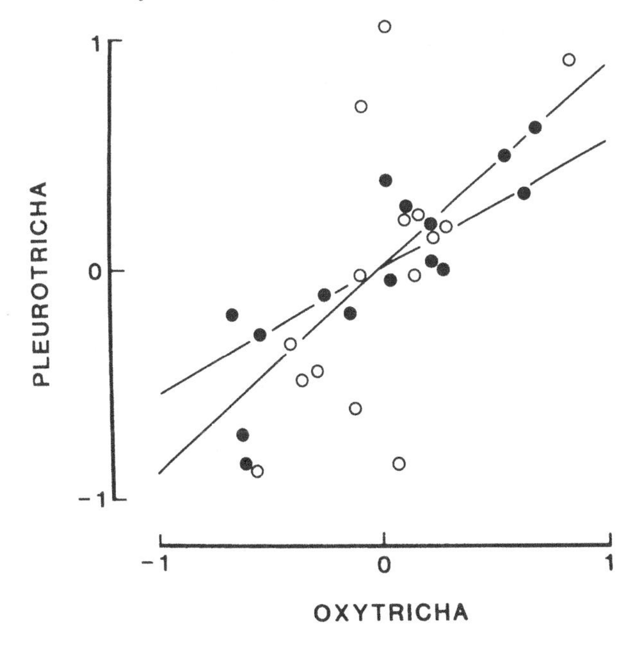

cence is then the result of growing organisms in conditions which are unable to support indefinite growth, and, as an immediate corollary, immortality would be achieved if only the right conditions were found. This is a strand which runs right through the history of the isolate culture experiments, beginning with Fabre-Domergue in 1889. As acute an observer as Jennings (1944a), remarked that although he began his observations using a concentrated infusion of lettuce, he switched to a dilute infusion when he found that his *Paramecium* grew just as well. But senescent decline set in much earlier in the dilute infusion, so he later switched back. The most striking demonstration of this phenomenon was an experiment with *Didinium* described by Beers (1933), which I have summarized in Figure 14. One series, cultured on well-fed *Paramecium*, maintained a high rate of fission throughout the brief (37-day) trial, whereas another group cultured on starved *Paramecium* rapidly senesced, and died out in fewer than 40 generations. However, 'senescent' animals removed from the latter group shortly before its extinction recovered the

Figure 14. Summary of Beers' experiment with *Didinium*. Symbols used are: solid circle, cultured on rich medium; open circle, cultured on poor medium; triangle, cultured on poor medium and then transferred to rich medium. Explanation in text. Source: Beers (1933).

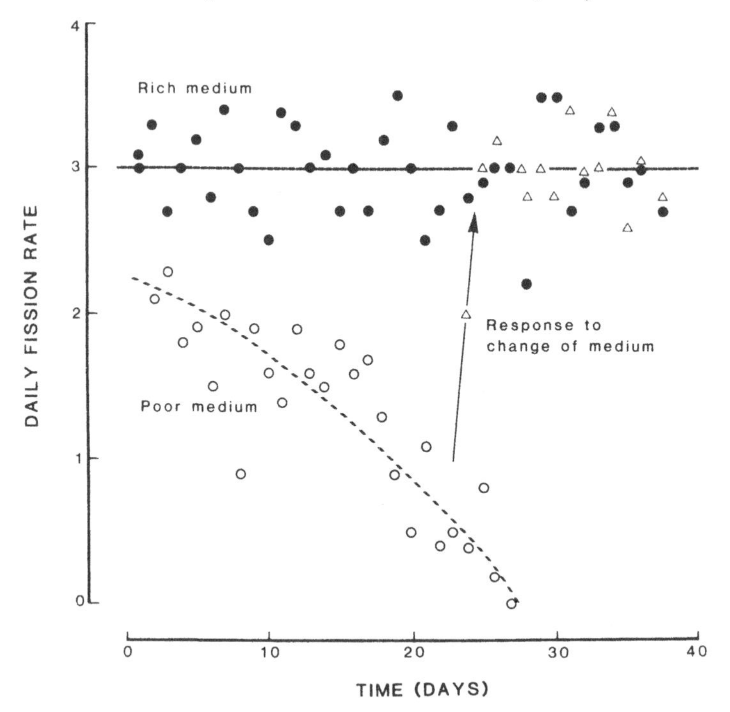

high fission rate of the former group within two days of being recultured on well-fed prey.

Beers' experiment provides the only satisfactory example of the rejuvenescence of failing cultures by a change of medium. This idea was first strongly advocated by Calkins (1902), who found that the declining division rate of *Paramecium* cultures could be stabilized and increased by transfer from hay infusion to meat extracts. After this procedure had worked several times it eventually failed, one of his stocks becoming extinct; a second stock, however, recovered after being treated with an extract of pancreas. This led Calkins to the conclusion that '...as the potential of vitality becomes reduced, something in the chemical composition of *Paramecium* wears out or disappears...a change of diet restores the power'. An indefinite lifespan in culture then 'seems to be merely a matter of finding the right food'. Other successful attempts to rejuvenate aging cultures were reported by Calkins and Lieb (1902) and Woodruff (1905). The possibility of such an effect cannot be dismissed out of hand; early workers allowed quite large numbers of cells to build up before transfer in their isolate lines, and their sterilization procedures were often inadequate, so that the long-term build-up of bacteria or bacterial products is not unlikely. However, the idea eventually fell out of favour because of the difficulty of obtaining consistent results. Calkins (1904) himself failed to rejuvenate senescent lines of *Paramecium caudatum*, and the determined attempts of Gregory (1909) Dawson (1926), and especially Austin (1927), to prolong the lifespan of ciliates such as *Uroleptus*, which normally show a rather predictable decline in vitality, were completely unsuccessful. Even worse, none of these failures could falsify the hypothesis, since it could always be claimed that the 'right food' had not yet been found. I do not know whether it was appreciated that the hypothesis was radically unsound because of its unfalsifiability, but, as with all hypotheses which fail to generate repeatable results under explicit protocols, interest in it gradually died away, and Beers' result is left as a solitary success.

The nature of any rejuvenescent effect of different medium is also unclear. Certainly it could not be attributed to the passive dilution of some essential nutrient, since isolate lines often flourished for 100 generations or more in the same medium, and a substance which is active at 2^{-100} of its original concentration is inconceivable. The drift of most speculation was more or less overtly Lamarckian, centering on the notion that a change of medium might produce heritable changes in the ciliates. This is not entirely absurd – the induction of heritable change by exposure to different nutrients is known for certain to occur occasionally in higher plants. There is, however, no suggestion in these cases that the induced change is appropriate – that it confers an advantage under the new conditions of growth. In protozoan rejuvenescence we are required to believe in the strictly Lamarckian proposition that on exposure to different conditions the induced change in the organisms is not only heritable but adaptive. Perhaps even this is not entirely out of the question. If there is any heritable response to altered conditions, it is possible that there is also additive genetic variance for the direction or, more likely, the magnitude of this response. Selection will then favour any change in the response which is advantageous under the new conditions, and the eventual result might be in line with the Lamarckian property of acquiring an appropriate heritable response to a specific environmental stimulus.

There is, of course, no evidence whatsoever that any such process occurred in the reported cases of rejuvenescence. Moreover, I think that a more conventional interpretation is quite satisfactory. Let us suppose that the decline in vitality is genetic, and is caused by the accumulation of a large number of mutations, each individually of small effect. Such mutations are quite likely to be rather specific in their effect on fitness: while causing a slight loss of vitality in some environments, they may actually elevate vitality in others. The potential importance of such interactions between genotype and environment is described later. For present purposes, the point is that the effect of a mutation depends on the conditions of culture. A collection of mutations which lower vitality under one set of conditions may be perfectly functional in others.

6

Does sex rejuvenate?

6.1 Sex and clonal longevity

The basic question of whether or not conjugation can extend the finite lifespan of a vegetative clone was settled by Calkins in 1919. Starting with a single individual of *Uroleptus mobilis*, he obtained exconjugants from crosses within the original parental line (clonal self-fertilization) or within or between the derived lines. These exconjugants were therefore rather highly inbred. Each line when propagated without conjugation had a finite lifespan of some 200–300 days, with a steep decline in fission rate during the latter part of this period (see Table 2 and Figure 5). The exconjugant lines, however, usually outlived their parent line, making it possible to maintain the organism in culture (Figure 15). Conjugation, therefore, restored a youthful fission rate and was essential for the continued existence of the stock. Figure 15 is one of the more important experimental results in biology, since it leaves no room for doubt that sex does indeed possess the rejuvenatory property so long claimed for it.

This result was confirmed by Woodruff and Spencer (1924), using *Spathidium spathula*, again a species which almost always shows a decline in fission rate culminating in extinction when cultured asexually (Table 2 and Figure 5). Like Calkins, they began with a single individual from which all their lines were ultimately derived. Their data are not quite straightforward to analyse, since many lines were discontinued before they became extinct; using only those lines which died out during the period of observation will therefore underestimate the true longevity of the derived lines, while using only those which were discarded before extinction will yield an overestimate. I have tried to summarize their results by calculating the difference between the longevity of a line and the subsequent longevity of the line from which it was derived, as a proportion of the sum of these two longevities; negative values thus indicate a decrease and positive values an increase in longevity due to conjugation. The results are give in Figure 16. Among the lines which died out, there is a large class of lethals which failed to divide at all; there are a few subvital clones which did not outlive the parental line, and then a second major mode comprising clones all of which outlived their parent. Among the clones which were discontinued, most outlived their parent. There is therefore a strong tendency for conjugation to produce lines which outlive a parental line

Figure 15. Rejuvenescence by conjugation in *Uroleptus*. Vertical lines indicate the lifespan of a line, the solid circle marking the point at which it became extinct. Time is indicated as generations on the left and as calendar time on the right. Source: Calkins (1919, 1920).

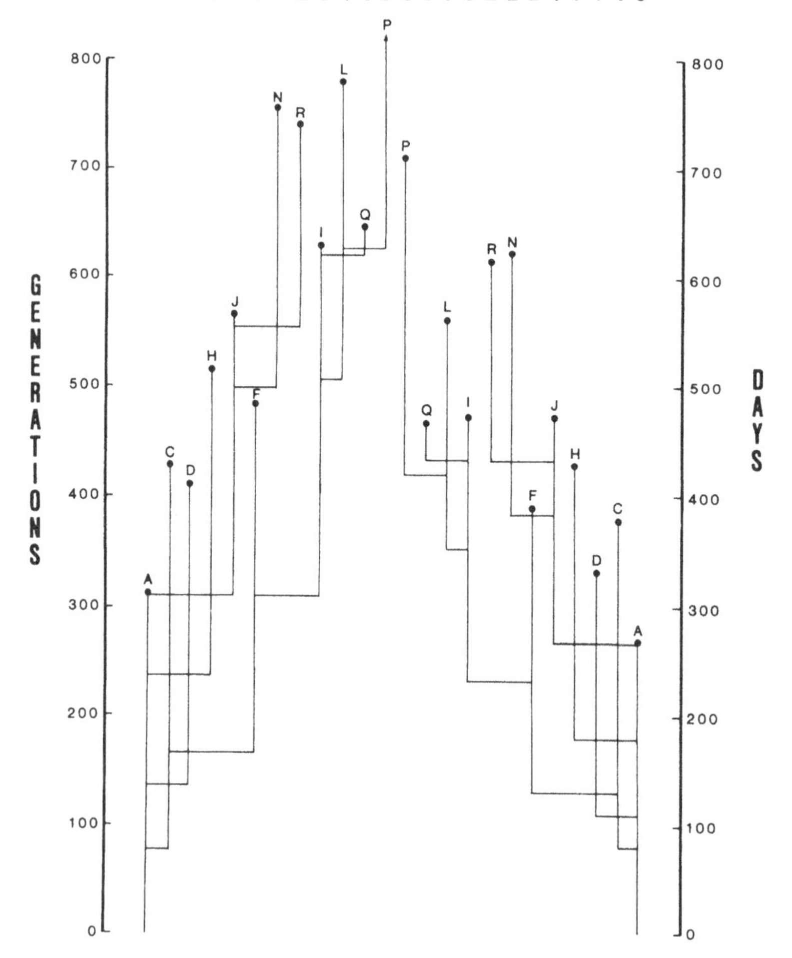

propagated asexually, not forgetting that a secondary result of conjugation is the production of many inviable lines. Among all the exconjugant lines, not fewer than 76/111 and not more than 90/111 outlived their parents, or would have done had they not been discontinued.

These experiments exclude any effect of automixis, which in *Uroleptus* and *Spathidium* occurs only in encysted individuals. There are, nonetheless, interesting complications arising from the age and degree of inbreeding of the conjugants, which are described below.

Figure 16. The effect of conjugation in *Spathidium*. Let: X be the lifespan of a culture, in generations; Y be the *subsequent* lifespan of the culture from which it was derived by conjugation; then the figure is a histogram of $(X-Y)/(X+Y)$. Cultures which died out during the period of observation are plotted above the line, cultures which were deliberately terminated below the line. A few cases in which both parents and exconjugants died immediately have been excluded. Source: Woodruff and Spencer (1924).

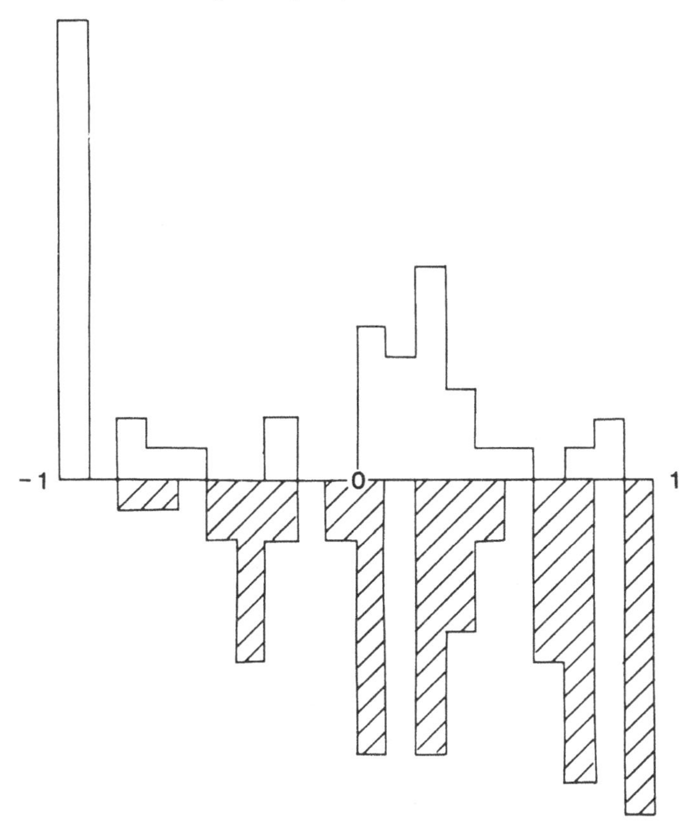

6.2 The inheritance of vitality

Interpreting the progressive loss of vitality during vegetative propagation and its restoration by sex in at least some protozoans requires some knowledge of the genetics of clonal longevity and fission rate – at the minimum, it requires an assurance that these are characters which can be transmitted from parent to offspring, both through sexual and through asexual reproduction.

For example, suppose that we have isolated a series of clones, obtained by crossing within some base population. We first want to know whether these clones differ appreciably with respect to fission rate. To do this, we must show that the variance of clonal means is substantial, when measured relative to the variance of fission rate within each clone. This requires raising several replicates of each clone, measuring the fission rate of each replicate, and then partitioning the total variance of fission rate into its within-clone and between-clone components. For example, Jennings created eight lines by conjugations within a single clone of *Paramecium aurelia*, and then measured the number of fissions during a five-day period in 16 replicates of each line (Jennings *et al.* 1932, Table 12, second period). A simple analysis of variance shows that the between-clone variance is indeed significant ($F_{7,120} = 16.8$, $P < 0.001$). Expressing the variance of clonal means as a fraction of the total variance yields the intraclass correlation coefficient, which in this case is $t = 0.496$. Since all of the replicates within a line are genetically identical, this implies that about half the variance in fission rate within the experimental group of clones is associated with genotypic differences. The nature of these differences cannot be ascertained from this sort of data; they may stem from the additive effects of different genes, or from interactions between allelic or nonallelic genes.

This is by no means an extreme result. Jennings (1913) employed the split-pair technique of Richard Hertwig. In the early stages of conjugation the two mating partners do not adhere very strongly, and can be separated by repeatedly drawing them into a fine pipette, before any exchange of cytoplasm or pronuclei. If these frustrated conjugants are then propagated asexually, we can partition the variance of fission rate between and within pairs of partners. Jennings' data yield $t = 0.045$, a very low value which shows that mating is essentially at random with respect to fission rate. On the other hand, if we allow conjugation to proceed normally and then isolate the exconjugants, we obtain the large and highly significant value of $t = 0.916$, suggesting that as much as 90% of the variance in fission rate may be genetically determined.

Granted the existence of extensive genetic variation in fission rate, we would like to confirm that differences in fission rate can be transmitted from parent to offspring. It would be surprising if this were not so in asexual reproduction, where any genetic elements concerned are presumably macronuclear. In the observations mentioned above, Jennings *et al.* (1932) in fact measured the fission rates of their lines during five successive periods. We can therefore ask whether there is any correlation between the mean value of clones at different times. The data yield $t =$ 0.724 (with $F_{7,31} = 13.8$, $P < 0.001$), showing that differences in performance are indeed perpetuated by asexual reproduction. This is not an isolated observation; several authors estimated the correlation between exconjugant and nonconjugant lines over successive periods of time,

Figure 17. Parent–offspring regression of fission rate, for conjugations within a clone of *Spathidium spathula*. The regression (solid line) is

$$y = 1.88 + 0.44x \quad (r^2 = 0.27)$$

for the absolute values. The broken line is the line of equality; exconjugants exceed nonconjugants for fission rates of less than 3.25, though the possibility that they exceed them over the whole range of the data cannot be excluded. Nonconjugants are the parental lines from which the exconjugants were derived.

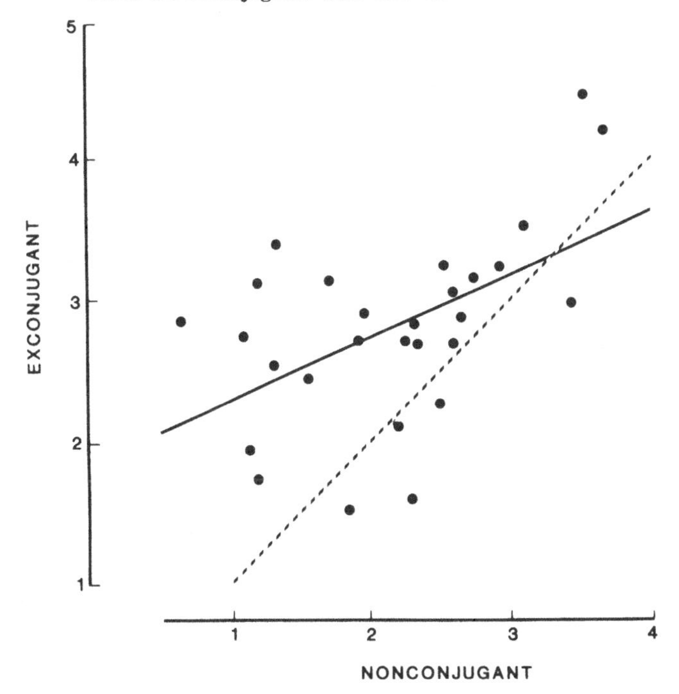

finding it to be about zero for nonconjugants but as high as 0.9 for exconjugants (e.g. Jennings 1913, Table 25; Cohen 1933).

The inheritance of fission rate during sexual reproduction is not very well documented; the heyday of cultural studies preceded the development of the modern theory of quantitative genetics, and consequently most of the reports are of little use. In particular, I have not been able to find any suitably detailed account of crosses between unrelated clones. The closest approach is made by studies which report the results of conjugation within a clone, of which the most extensive is that described by Woodruff and Spencer (1924) for *Spathidium spathula*. Their data yield a value of 0.44 for the regression of exconjugant on nonconjugant, which is shown in Figure 17. This value represents a heritability, but cannot be interpreted in any simple way (except as demonstrating the existence of some heritable genetic variance) since the parental and derived lines were kept in different environments.

Clonal longevity is also transmitted through conjugation, indicating an effect of the micronuclear genome. Jennings (1944c) made crosses between six clones of *Paramecium aurelia*, each of which had differentiated into compatible mating types. Although not all combinations of clones and mating types were assayed (so that a full diallel analysis is not possible), we can make crude estimates of heritability by treating his data as blocks of half-sib and full-sib families. The raw data are the fractions of lines dying out from each cross. These can be converted into liabilities, assuming an underlying normal distribution of the factors contributing to clonal mortality. Treating the liability for any given cross as an estimate of the probability that a random exconjugant from that cross would die, we can then arrange Jennings' data into 6 half-sib families or 11 full-sib families. The intraclass correlation coefficient is very high in both cases: $t = 0.329$ for half-sibs (where it estimates one-quarter of the additive variance), and $t = 0.859$ for full-sib families (where it estimates half the additive variance, plus one-quarter of the dominance variance). Perhaps because of the small number of families, the between-family effect is not formally significant for half-sibs ($F_{5,14} = 2.59$, $P \approx 0.20$), but it is highly significant for full sibs ($F_{10,23} = 19.7$, $P < 0.001$).

Differences in fission rate and clonal longevity are, then, transmitted from generation to generation, through macronuclear elements during asexual reproduction and through micronuclear elements during conjugation. The heritability of both characters appears to be substantial.

The effects of conjugation on the mean and variance of fission rate within a line are much better-defined, although again the only detailed data sets refer to clonal self-fertilization. One typical result, from Cohen's work on *Euplotes*, is shown in Figure 18(*a*). The daily fission rates of

Figure 18. Effect of conjugation on variability. Both sets of diagrams give the frequency distribution of the number of fissions per five days, as deviations from the unweighted joint mean, of exconjugants and nonconjugants. (*a*) *Euplotes patella*. Conjugating and nonconjugating individuals taken from the same mass culture, founded by a single individual. Source: Cohen (1933). (*b*) *Paramecium aurelia*. Selfing within six clones. Source: Jennings *et al.* (1932), Table 5.

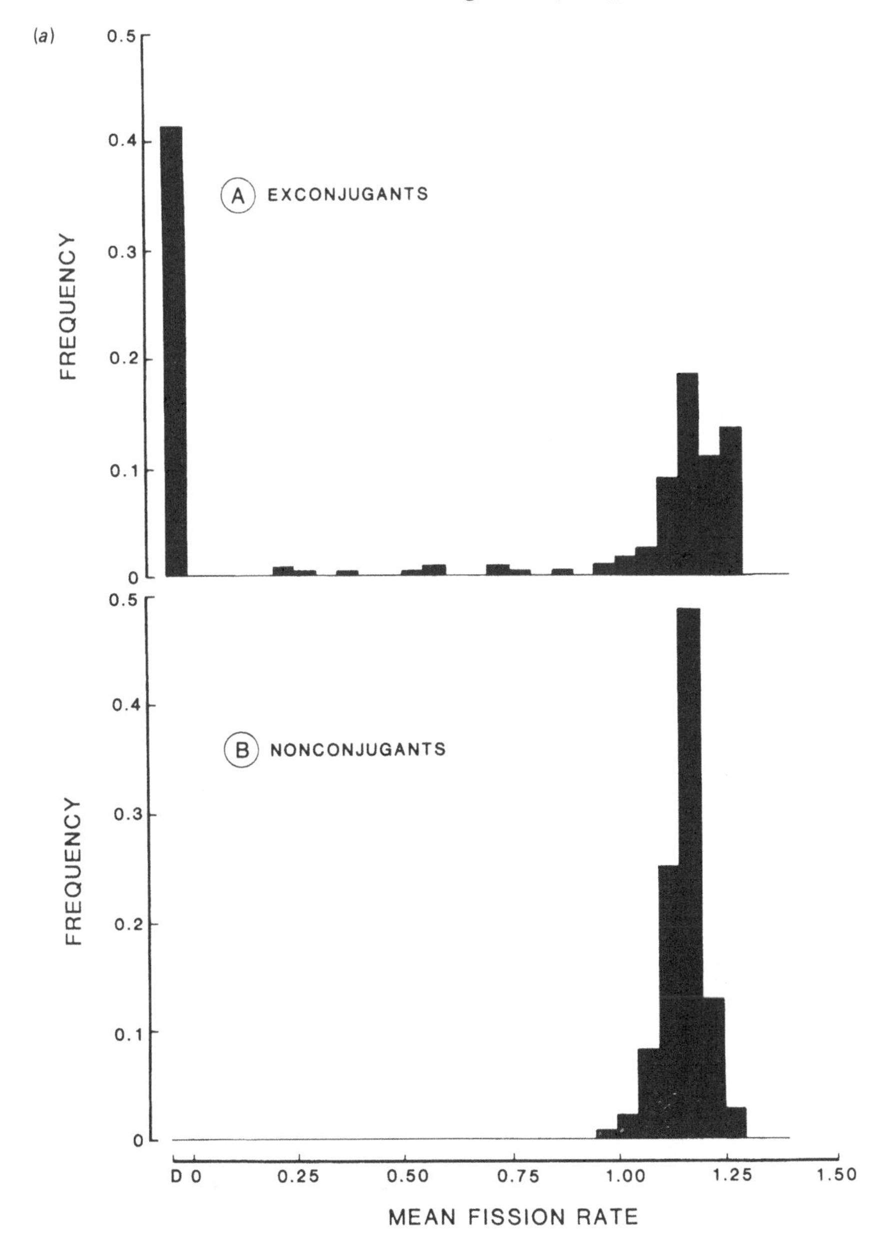

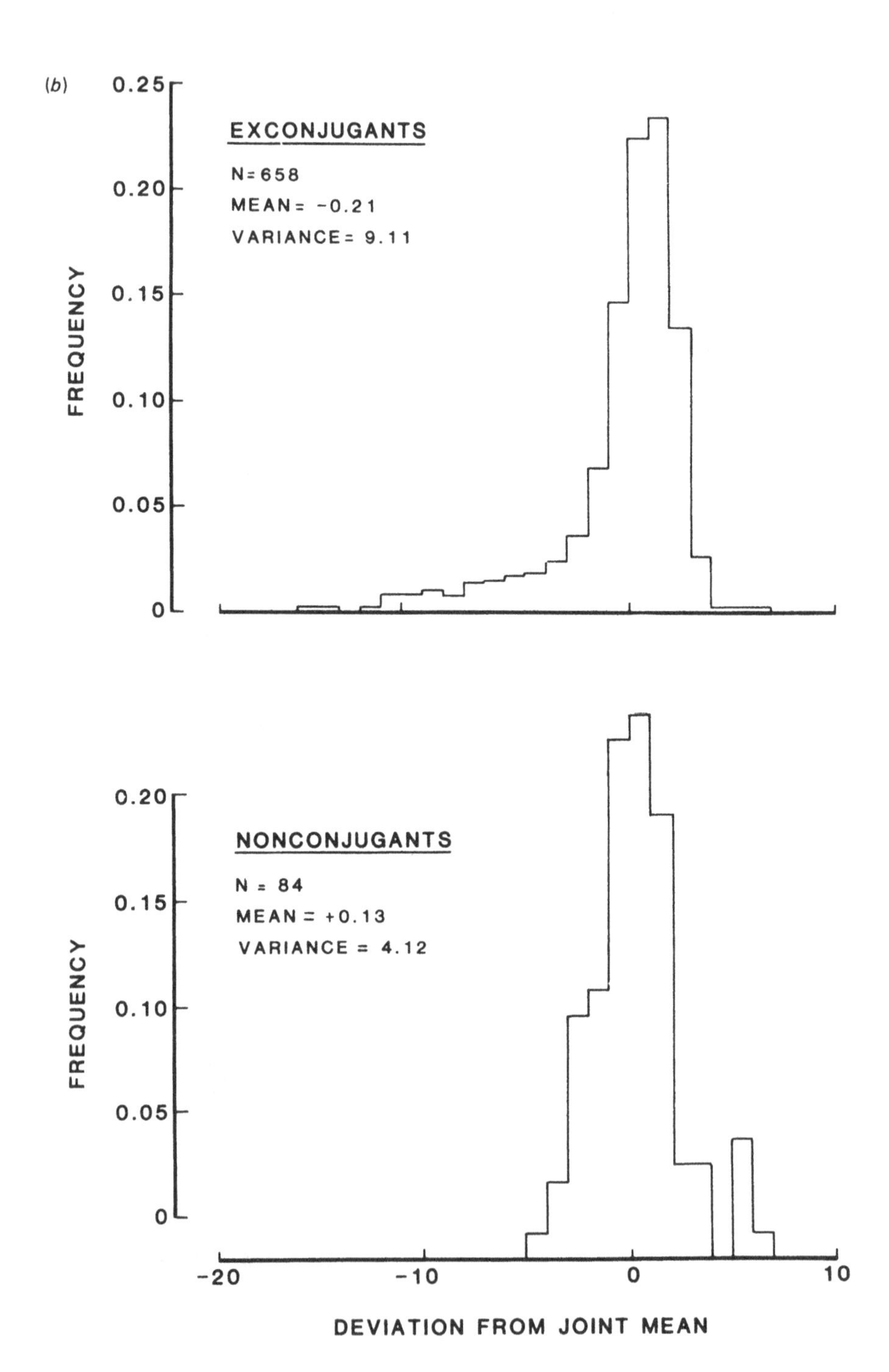
(b)
EXCONJUGANTS
N=658
MEAN= -0.21
VARIANCE= 9.11
FREQUENCY
0.25
0.20
0.15
0.10
0.05
0
NONCONJUGANTS
N = 84
MEAN = +0.13
VARIANCE = 4.12
FREQUENCY
0.20
0.15
0.10
0.05
0
-20
-10
0
10
DEVIATION FROM JOINT MEAN

Figure 19. The effect of conjugation on the mean and variance of fission rate. Symbols plotted are: solid circle *Paramecium aurelia* (Sonneborn and Lynch 1937); open circle *Paramecium aurelia* and *Paramecium caudatum* (Jennings 1913); triangle *Euplotes patella* (Cohen 1933); cross *Paramecium aurelia* (Raffel 1930). The solid line is the line of equality.

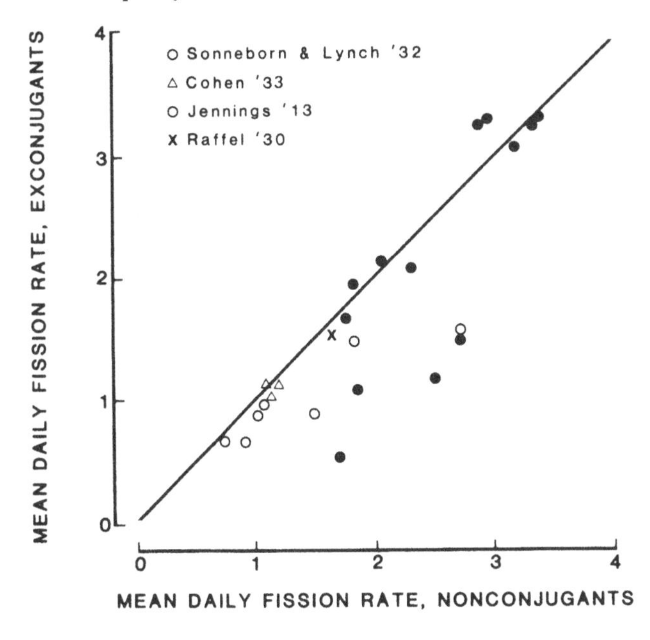

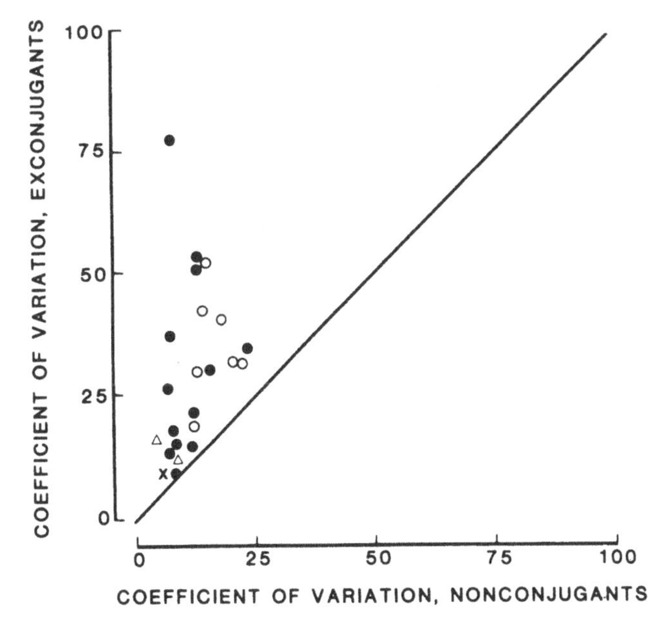

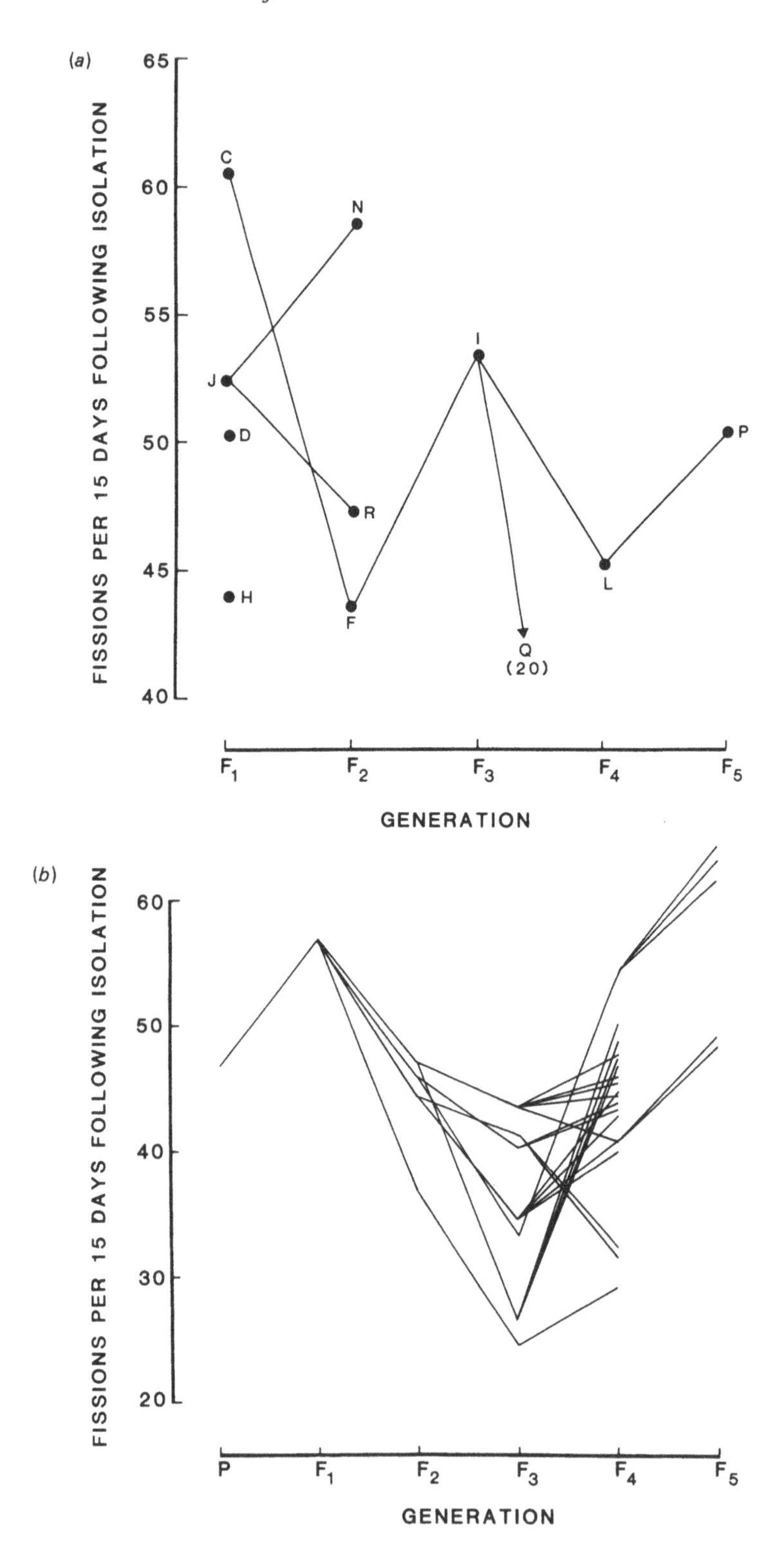

(a)
65
60
55
50
45
40
FISSIONS PER 15 DAYS FOLLOWING ISOLATION
C
N
I
J
D
R
H
F
L
P
Q
(20)
F1
F2
F3
F4
F5
GENERATION
(b)
60
50
40
30
20
FISSIONS PER 15 DAYS FOLLOWING ISOLATION
P
F1
F2
F3
F4
F5
GENERATION

nonconjugants are distributed more or less normally within a narrow band between about 1 and 1.25. The exconjugant distribution is bimodal, with one mode at lethality and another mode near to the nonconjugant mode, the two modes being connected by a series of subvital lines. A similar result was obtained from conjugations within six *Paramecium* clones by Jennings (1932), except that there were no lethal lines; instead, the exconjugant and nonconjugant distributions are similar except for the negative skew of the exconjugants, created by a long tail of subvital lines (Figure 18(*b*)).

Results like these suggest that conjugation, or at least clonal self-fertilization, has little effect on the mean fission rate, though usually depressing it slightly, whereas the variance is markedly increased. The generality of this conclusion is shown by figure 19.

Finally, one might ask whether or not any effect on the mean fission rate is cumulative, so that successive conjugations cause a steady increase or decrease in the mean, relative to the original stock. Figure 20 shows that this is not the case. Though there are wide fluctuations, conjugation does not procure any steady improvement in performance.

6.3 The effect of inbreeding

The harmful effects of close inbreeding were recognised by several early authors, including Butschli and Maupas, though the evidence remained very sketchy until Jennings' experiments. In some cases, such as *Blepharisma undulans* (Calkins 1912) and *Stylonichia pustulata* (Baitsell 1912), conjugation between members of the same clone invariably resulted in the death of the exconjugants without division. In others the situation was less extreme; in *Paramecium aurelia*, for example, Calkins (1902) found that 5/6 exconjugants survived from matings between random individuals from pond samples, whereas only 5/80 exconjugants survived from various clonal self-fertilizations. Jennings (1944c) studied reciprocal matings within and between six defined *Paramecium aurelia* clones. His results, which are summarized in Figure 21, show that selfing is normally, but by no means always, associated with a greater mortality of exconjugants. In 31 comparisons between the mortality of a cross and the mean mortality of the two clones when selfed, the mortality of lines from

Figure 20. The effect of successive conjugations. (*a*) *Uroleptus* (Calkins 1919). Letters indicate different series; when the first (C) exconjugant series was derived the parental line was averaging 55–60 divisions per 30 days. (*b*) *Spathidium* (Woodruff and Spencer 1924). All lines completing four or more sexual generations are plotted.

the cross was lower in 22 cases. Perhaps more revealing is the fact that if we take any given clone, the mortality of exconjugants produced by outcrossing is related to the mortality produced by selfing. Lines which have a very low mortality on selfing have somewhat greater but still low mortality when crossed; those which suffer high mortality on selfing have a somewhat lower mortality when crossed. For a mortality of about 30% of all exconjugant lines, crossing and selfing were equivalent.

These results were obtained from compatible self-differentiated lines held in mass culture and mated at intervals, the conjugants descending vegetatively from the lines which yielded the previous set of conjugations. This amounts to clonal self-fertilization. Jennings distinguishes it from another scheme, which he calls 'sib-mating'. This scheme involves mating individuals which descend vegetatively from the surviving exconjugants of

Figure 21. Mortality in selfed and crossed conjugations of *Paramecium aurelia*. Plotted points are mortality among exconjugants from crosses plotted at mortality of same line when selfed. The different points for a given cross, joined by a vertical line, are for different combinations of mating types. The horizontal line, which is drawn through the mean over all combinations of mating types for a given cross, indicates the range of values for selfed lines. The eight crosses shown involve 6 distinct clones. The broken line is the line of equality. Source: Jennings (1944c), Tables 4–13.

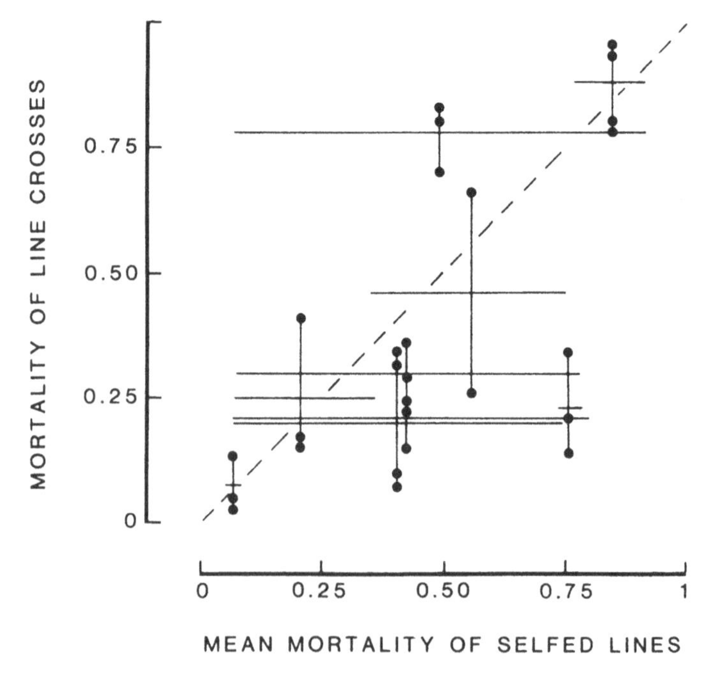

a previous mating. The difference between sib-mating and clonal selfing lies solely in the length of the independent vegetative ancestry of the conjugants. In clonal selfing the two conjugants may be hundreds or even thousands of generations removed from the common ancestor of their two lines; in sib-mating only a few dozen fissions need pass before the exconjugant lines acquire sexual competence and can be remated. Consequently, sib-mated individuals will be identical at virtually all their loci, whereas clonal selfing gives much more opportunity for the conjugants to have diverged genetically by mutation. The difference in the

Figure 22. The effect of repeated close inbreeding in *Paramecium*. The symbols refer to different crosses: open circle BH87 (self-fertilized); half-open circle, BH89 × BH90 (two clones from the same locality); solid circle BS stock (founded by two clones from distant localities). Because of age effects, only results from clones less than 10 months old are shown. Source: Jennings (1944c), Table 4, 6 and 8.

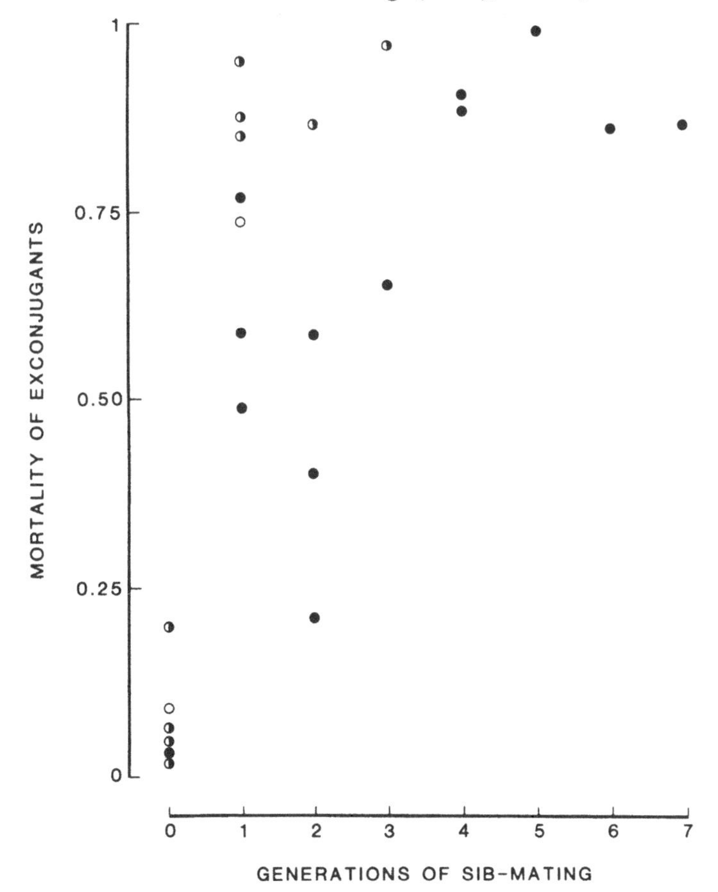

mortality of exconjugants between the two schemes is dramatic. Sib-mating results immediately in a very high fraction of lethal exconjugants, and if several generations of sib-mating are practised in succession this fraction rises asymptotically towards unity (Figure 22).

The simplest explanation for the mortality associated with extreme inbreeding is in terms of the segregation of recessive lethal genes arising by mutation after the isolation of the founding individual. In clonal selfing, the two conjugants have long separate histories, and recessive lethals are likely to arise at different loci; exconjugant mortality will therefore be relatively low. In sib-mating, on the other hand, the two conjugants are identical and therefore bear recessive lethals at the same loci. Suppose that the genome of the sib-mated individuals bears a recessive lethal at a certain locus. The probability that this will be carried by the pronuclei created by meiosis in one of the partners is $\frac{1}{2}$; similarly,

Figure 23. Age and inbreeding. The mortality of outcrossed (BH89 × BH90, solid circle) and sib-mated (BS, open circle) exconjugants at different ages. Source: Jennings (1944c), Table 6 and 8.

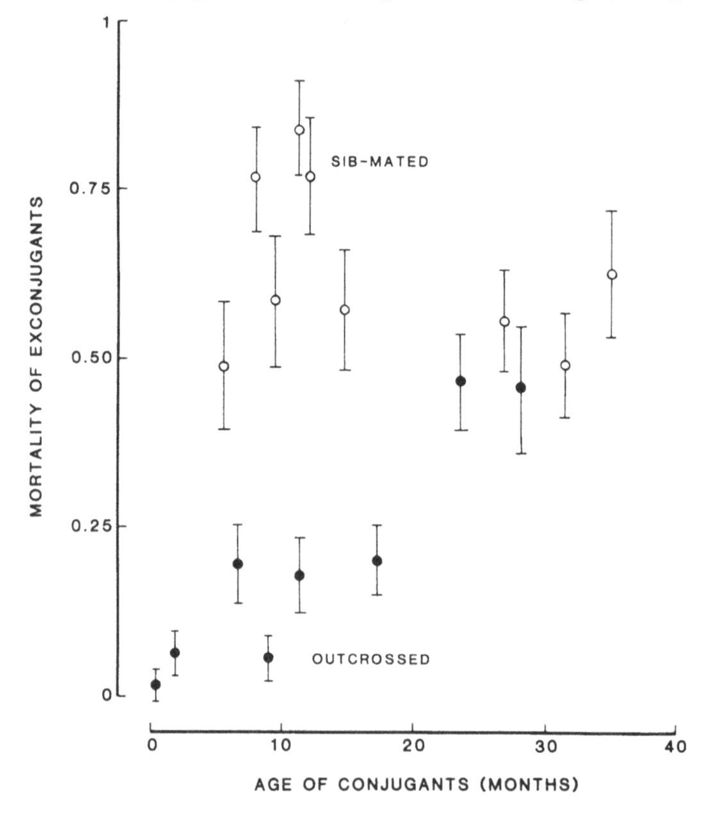

the probability that it will be transmitted by the other partner is $\frac{1}{2}$. Therefore, $\frac{1}{2} \times \frac{1}{2} = \frac{1}{4}$ of exconjugant lines will be homozygous for the lethal, and will die out before dividing. More generally, if recessive lethals occur at n independent loci, the fraction of exconjugants which are homozygous for at least one lethal is $1-(\frac{3}{4})^n$. Thus, the fraction of lethal lines generated by sib-mating increases from 0.25 when only a single recessive lethal is segregating, up to about 0.99 for 15 lethals per diploid genome. In the largest data set reported by Jennings (1944c, Table 8, mating BS2), the fraction of lethal lines in the first generation of sib-mating was 265/1098 = 0.241. This suggests the presence of about one recessive lethal per diploid genome, which is in line with estimates from other organisms. Successive sib-matings cause a progressive rise in mortality because the mth sib-mating renders the conjugants identical for all the recessive lethals which have arisen in the previous $m-1$ cycles.

6.4 Conjugation in aging clones

Although Maupas believed that conjugation caused rejuvenescence, he stressed that the cells had to be the right age: if they were too young they would not conjugate, while if they were too old most of the exconjugants died. For example, I have stressed above the increased mortality associated with close inbreeding, illustrating this with data from Jennings' experiments. But Figure 23 shows that the situation is complicated by the effect of clonal age. The mortality of sib-mated lines is indeed much greater than that of outcrossed lines, provided that both are relatively young. In fact, the mortality associated with sib-mating does not seem to change appreciably with clonal age. The mortality of outcrossed exconjugants, on the other hand, increases with the age of the conjugants to such an extent that after about two years in culture (i.e. some thousand generations of asexual fission) out-crossing and sib-mating give about the same proportion of lethal exconjugants.

Although several earlier (and later) authors describe age effects, Jennings' work is by far the most extensive, and I shall confine myself to summarizing his results. The basic phenomenon is shown in Figure 24, which shows that the mortality of exconjugants rises with the combined age of the conjugants. The effect is a large one, with average mortality rising from less than 10% when both conjugants are very young towards 100% when the sum of their ages exceeds 100 months (or some 2000+ asexual generations). There is no doubt of the reality of this effect – Figure 24 is based on some 10000 exconjugant lines from over 150 crosses – but there is also a good deal of scatter around the main trend. This variance comes mainly from two sources. The first is simply the variance arising from the differences between clones, which can be appreciated from the different symbols plotted in Figure 24. In Figure 25 I have used another data set to plot several clones separately, showing that exconjugant mortality increases with parental age in most or all of them, but at different rates and possibly to different levels. This phenomenon is not peculiar to *Paramecium*. For example, Frankel (1973) found that the mortality of exconjugant lines in the unrelated hypotrich *Euplotes* rose from 1–2% to about 95% after 700 generations of fission.

The second source of variance involves an interaction between the ages of the conjugants. I have illustrated this in Figure 26 by plotting the sum of parental ages on their difference, indicating the mortality of each cross. When both conjugants are young, mortality is almost always low. It increases with the combined age of the conjugants, but it is not noticeably greater when both are old than when only one is old. For moderate values

Figure 24. The relationship between the mortality of exconjugants and the age of the conjugants. 'Mortality' is the fraction of exconjugant lines dying out within 5 generations. Age of conjugants is given as the combined age in months of the two conjugants, either from collection or from the last conjugation. The number of exconjugant lines tested varied from 12 to 266, and was usually 48–96. Letters refer to crosses involving the same clone: A clone 2; B clone 6; C clone 12; D clone 35; solid circles are a variety of other clones. Asterisks are mean values of mortality at 10-month intervals of combined parental age. The regression of the means is

$$y = 0.082 + 0.00854x \quad (r^2 = 0.74)$$

Source: Jennings (1944c).

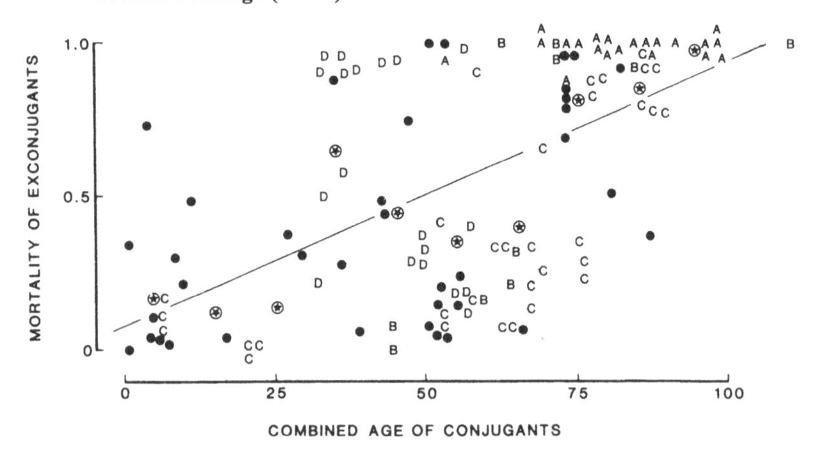

of the combined age, therefore, mortality is greater when one clone is young and the other old than when both are about the same age. The results of these crosses are given numerically in Table 4, whose most striking feature is the very high mortality of crosses involving one very old cell, regardless of the age of its partner.

The harmful effects of clonal age bear fundamentally the same interpretation as the harmful effects of inbreeding, namely the accumulation of harmful or lethal recessive mutations. The older the clones involved in a cross, the greater is the probability that the same recessive mutation has occurred in both. The interaction between the ages of conjugants cannot be explained in the same way. Suppose that two conjugants bear respectively m and n mutations out of a total of Q loci, where $m+n =$ some fixed number C. Then the probability that a random exconjugant is homozygous for at least one mutation is

$$1-\sum_{i=0}^{n-1}[1-m/(Q-i)] \approx 1-\exp[-(C-n)\sum_{0}^{n-1}(Q-i)^{-1}]$$

which is an increasing function of n; there will be a lower probability of obtaining a recessive homozygote when one conjugant bears many mutations and the other few than when both bear about the same number,

Figure 25. Effect of clonal age on mortality of exconjugants. This data is independent of that plotted in Figure 24, with which it should be compared. Lines connect points from the same cross at different ages of the parental clones. Source: Jennings (1944c), Table 1 and 8.

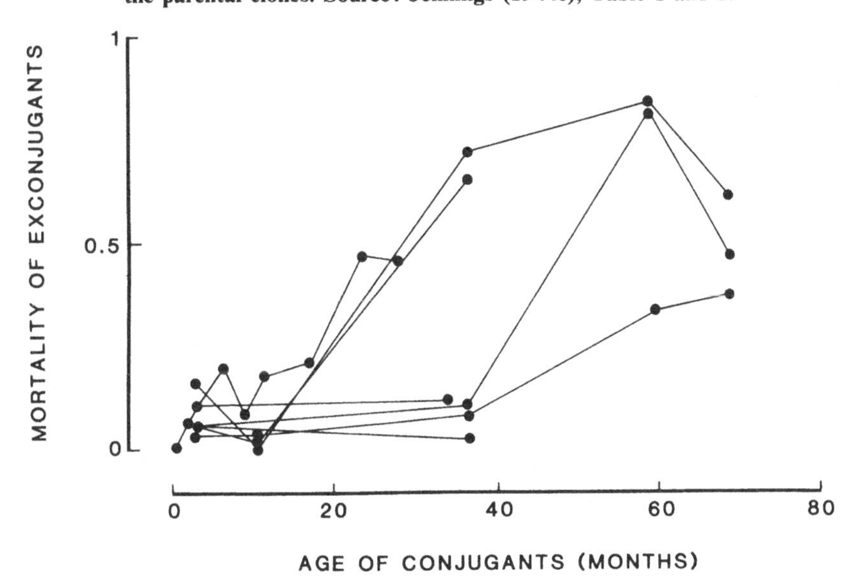

Figure 26. The longevity of exconjugants as a function of the ages of both conjugants. The vertical axis is the sum of ages in months, as in Figures 23 and 24; the horizontal axis is the difference between their ages. The solid line is the line of equality; by definition, all points must fall above this line, unless one clone is age zero, in which case the point will fall on the line. Thus, points at top right represent crosses between young and old clones; at top left, between two old clones; and at bottom left, between two young clones. The symbols indicate the mortality of the exconjugants: solid circle 90%, half-open circle 50–90%, open circle 10–50%, cross less than 10%. Source: Jennings (1944b).

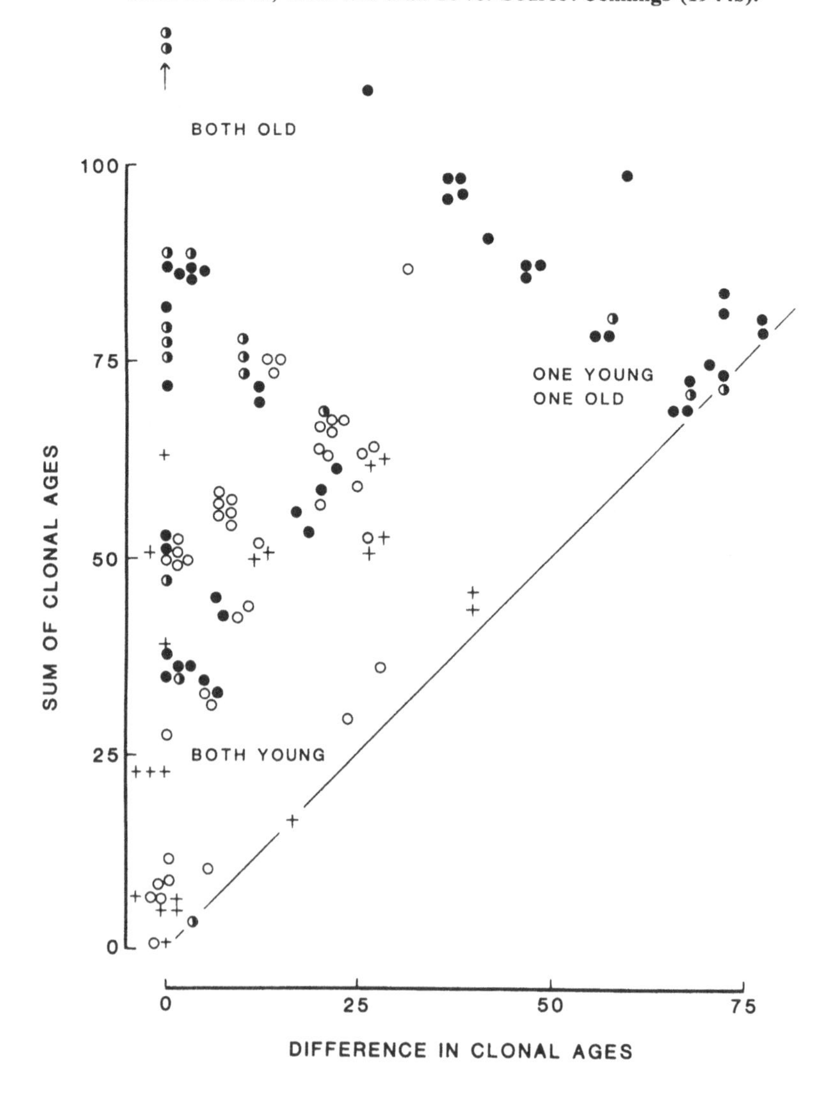

Table 4 *The mortality of exconjugant lines as a function of the ages of the clones crossed*

The upper number in each cell is the fraction of lines failing to complete five successive fissions. The lower numbers, in parentheses, are sample sizes: the number of crosses, followed by the number of exconjugant lines observed. The total sample comprises 157 crosses (among 44 available clones) and 10556 exconjugant lines. Digested from Jennings' 1944 papers.

Age of younger clone	Age of older clone (in months): 0–10	10–20	20–30	30–50	50+
0–10	0.286 (17,1248)	0.041 (2,96)	0.312 (2,96)	0.205 (3,156)	0.959 (2,1300)
10–20		0.474 (14,1186)	0.753 (7,405)	0.278 (17,1148)	0.828 (10,338)
20–30			0.557 (10,1051)	0.269 (14,1038)	0.843 (14,492)
30–50				0.791 (21,1826)	1 (3,72)
50+					0.730 (2,204)

for any given total load. Nor can dominant mutations be responsible, since the probability that one has arisen in either conjugant will be simply proportional to the sum of the ages of the conjugants. A cytoplasmic effect is scarcely more plausible, not only because of the small quantity or cytoplasm transferred during conjugation but also because it is difficult to understand why a cytoplasm immediately toxic to both exconjugants should not have been toxic to the older exconjugant.

6.5 Can automixis rejuvenate?

Although Woodruff's *Paramecium aurelia* went through several thousands of generations in isolate culture without conjugation, its members underwent a periodic nuclear reorganisation involving the division of the micronucleus and the breakdown and restoration of macronucleus. Originally described as 'endomixis', and interpreted as a mitotic division of the micronucleus accompanied by macronuclear regeneration, this process was almost certainly automictic (Diller 1936), with a meiosis followed by the fusion of sister pronuclei creating a completely homozygous micronucleus (Sonneborn 1954) as the template for the new macronucleus. This is, so to speak, the limiting case of inbreeding.

The immediate effect of automixis is to depress vitality by increasing the death-rate. This may be due in part to the metabolic stress of major intracellular change, but since the effect persists for ten generations or so after automixis has begun (see Caldwell 1933) a genetic hypothesis seems more reasonable. A very striking fact is that the effect of clonal age on the mortality of exconjugants is parallelled by its effect on the mortality of exautomicts. In this case, clonal age is the time which has passed since the last automixis, in non-conjugating stocks, and as it increases the mortality of the exautomicts increases to nearly 100% (Figure 27).

Unfortunately, there has been no study of the effects of automixis on variation comparable with Jennings' work on conjugation. Erdmann (1920) attached great importance to variation in size appearing at automixis, but used the quite inappropriate technique of culturing and comparing the lines derived by fission from a single exautomict. More reasonable procedures were used by Sonneborn and Lynch (1937) to show that exautomicts could be quite different from their parental lines, and by Caldwell (1933) to show that the exautomicts could differ from one another, although neither paper is very extensive. In Figure 28 I have shown how Caldwell's two lines, each founded from a single individual, increased in variability through time as they underwent a series of

Figure 27. The relationship between mortality and the length of the interautomictic period. Solid circles are data from Pierson (1938), open circles data from Gelber (1938).

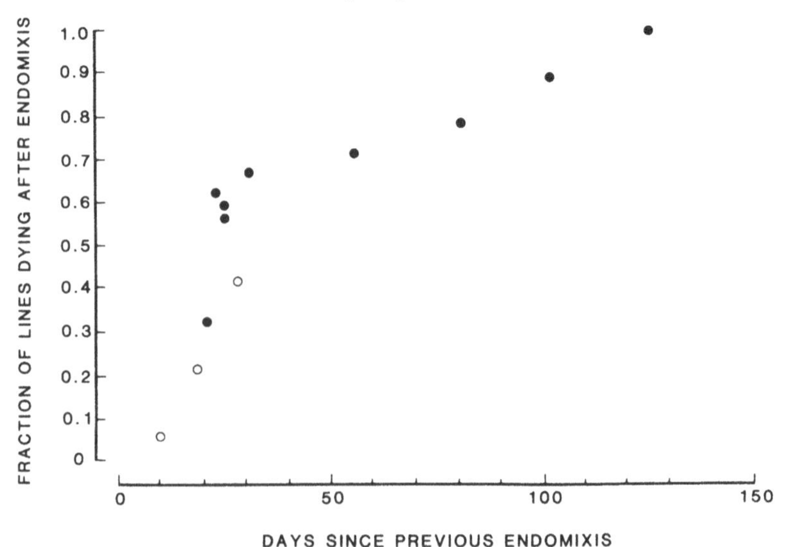

automixes. This is presumably due to the segregation of homozygotes from an originally heterozygous cell; the trends cannot be taken quite at face value, since Caldwell deliberately selected extreme lines for further cultivation, but they do clearly demonstrate the generation of heritable variance in fission rate through automixis.

The initial description of automixis (as endomixis) was closely tied up with the analysis of short-term 'rhythms' in the fission-rate of long-lived series of *Paramecium aurelia*. It was claimed by Woodruff and his collaborators that automixis occurred after a period of declining fission rates, and had the effect of restoring fission rate to its original level. The rhythms, whose validity I have established above, are therefore generated by the periodic arrest of senescence by automixis. This claim was supported by fission-rate histograms in which the rough coincidence of automixis with minimal rates of fission was taken to be self-evident. In Table 5 I have reanalysed this data more objectively, and come to the same conclusion. For each five-day period which includes an automixis I have extracted the values of fission rate for the two five-day periods on either

Figure 28. The increase in the variability of fission rate in stocks undergoing automixis. The coefficient of variation of daily fission rate between lines is given at different times following the initial isolation of the series. Source: Caldwell (1933), Table 2. Different symbols refer to two different clones.

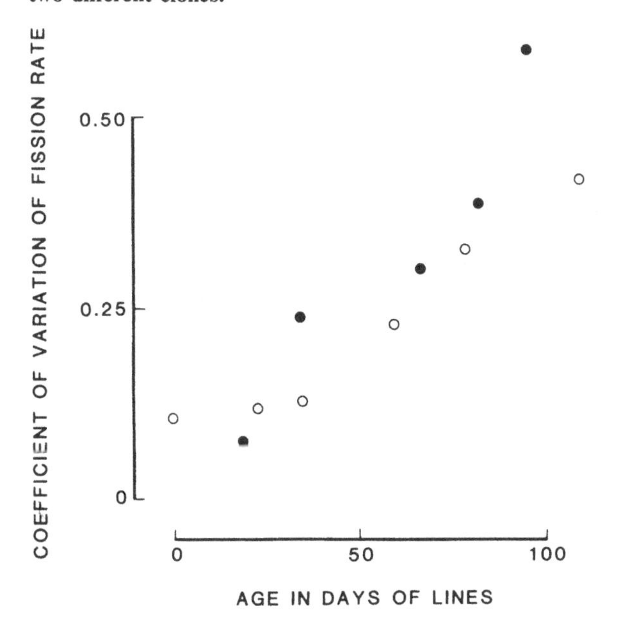

Table 5. *Fission rate before and after automixis*

Automixis occurs at time t, each time period being five days; raw data are mean fission rates over each of the five periods between $t-2$ and $t+2$. F/F_{t-2} is the fission rate at a given time, relative to the fission rate at time $t-2$. $\Delta F/F$ is the difference between fission rates in successive periods of time, divided by their mean value; thus, $\Delta F/F$ for period $t+1$ is $2(F_{t+1}-F_t)/(F_{t+1}+F_t)$. Values given are mean ± 1 s.e.; for F/F_{t-2} the statistics are back-transformed from logarithms. Each value is based on about 60 observations, intrapolated from histograms. Sources: Erdmann and Woodruff (1916), Fig. 6; Woodruff and Erdmann (1914), Fig. 16; Woodruff (1917a), Figs. 1 and 2; Woodruff (1917b), Figs. 1–5.

Time	F/F_{t-2}	$\Delta F/F$
$t-1$	0.9042 ± 0.0417	-0.1018 ± 0.0373
t	0.7283 ± 0.0409	-0.2045 ± 0.0305
$t+1$	0.9507 ± 0.0585	$+0.2657 \pm 0.0373$
$t+2$	0.9701 ± 0.0648	$+0.0196 \pm 0.0270$

Figure 29. Senescent decline in the absence of autogamy. The thick barred line is the mean fission rate in a series of daily isolate lines, in which autogamy occurred about every 15 days. The two thin lines are two lines derived from this autogamous stock, from which all autogamous individuals were removed and replaced with nonautogamous sisters. Source: Sonneborn (1954), Table 1, series C, P and 0–20.

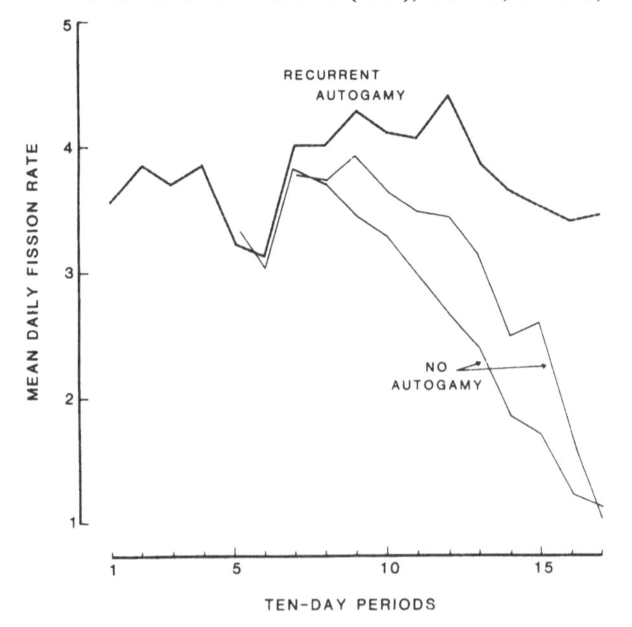

side. Thus if the mean rate of fission during the five-day period in which automixis was detected is F_t, I have intrapolated not only F_t but also F_{t-2} and F_{t-1}, and F_{t+1} and F_{t+2}, from the published histograms. About 60 such sets of values can be extracted from the publications of Woodruff and Erdmann. If we express all values relative to F_{t-2}, well before automixis occurs, we can see that F_{t-1} is generally less than F_{t-2}, while F_t is still smaller. The low value of F_t might be attributable in part to the directly disruptive effect of automictic reorganisation, but this cannot be the case for F_{t-1}. Following automixis, fission rate recovers, probably completely; mean values of F_{t+1} and F_{t+2} are not significantly different from F_{t-2} or from one another, although both are slightly less than F_{t-2}. Calculating the rate of change ΔF relative to the mean fission rate at any given time leads to the same conclusion, as it should. $\Delta F/F$ is negative before automixis, positive in the first period following automixis, and zero in the next period. Woodruff's contention that automixis follows a period of declining fission rates and results in rejuvenescence therefore seems to be sound.

Direct evidence for rejuvenescence, however, was not obtained until much later, though Calkins (1915) described rejuvenescence among individuals emerging from cysts in *Didinium*, a ciliate in which automixis occurs only during encystment. The crucial experiment was performed by Sonneborn (1954), who compared control lines in which automixis regularly occurred with derivative lines in which it was prevented by the removal of automictic individuals and their replacement by nonautomictic sisters. While the automictic lines showed no trend in fission rate with time, fission rate in their nonautomictic derivatives almost invariably decreased. Two of his largest data sets are illustrated in Figure 29. The automictic series had an apparently indefinite lifespan, though it was not cultured beyond a few hundred generations; without automixis, progressive senescence resulted in extinction after a greater or lesser number of generations, the clones differing in their longevities.

7

Germinal senescence in multicellular organisms

Multicellular animals and plants are larger and longer-lived than protists, and long-term isolate culture demands a great deal of patience. There are very few cases in which asexual metazoans or metaphytes have been taken through more than a hundred generations in the laboratory. The available literature is summarized in Table 6.

7.1 Plants

I have already mentioned Hartmann's culture of the colonial green alga *Eudorina*, which went through over 228 generations with no loss of vigour (see Table 2). Uspenski and Uspenskaya (1927) cultured the related *Volvox aureus* through 47 asexual generations in 15 months, but do not give details of the rate of reproduction. These algae are haplonts.

A particularly vivid instance of senescence and rejuvenation was described by Mather and Jinks (1958) in the fungus *Aspergillus*. Continued propagation by asexual spores led to a rather steep decline in perithecial production within 10 or so generations, but vitality was completely restored by a single sexual episode. This seems to implicate a cytoplasmic deterioration analogous to macronuclear senescence in ciliates.

Although vascular plants are generally too long-lived for isolate culture to be feasible, it is a commonplace of horticulture that some vegetative clones are very stable and can be transmitted by tubers (e.g. potatoes) or grafts (e.g. pear-trees) for many generations without deterioration. There is, of course, powerful artificial selection against any inferior lines that do emerge. More generally, the aging of meristems is also a commonplace, in terms of the size, shape and physiology of leaves borne at successive nodes. However, the only laboratory study of meristematic aging concerns the greatly-reduced aquatic form *Lemna* (Ashby *et al.* 1948). This

Table 6. *Longevity of pedigreed cultures of asexual multicellular organisms*

The data columns are: REPN manner of reproduction (egg, by eggs, apomictic algae, rotifers and cladocerans; veg, vegetative, architomical budding in hydras, more nearly paratomical budding or fission in duckweeds, turbellarians and oligochaetes).
MAXG greatest recorded number of consecutive asexual generations.
MAXT approximate duration of longest-lived cultures in days.
SENX whether or not senescent decline in vitality was observed; yes/no means that results vary according to whether clones were developed from early or late offspring (rotifers) or from anterior or posterior fragments (turbellarians and oligochaetes).

Note that several reports of much greater clonal longevities, especially for hydras, turbellarians and fungi, have been excluded because of the vagueness of the sources.

Organism	REPN	MAXG	MAXT	SENX	Authority
Chlorophyte algae					
Eudorina	egg	228+	1900+	no	Hartmann 1921
Volvox	egg	47	450	no	Uspenski and Upenskaya 1927
Ascomycete fungi					
Neurospora	egg	14		yes	Mather and Jinks, 1958
Angiosperms					
Lemna	veg	c10	c400	no	Ashby and Wangermann 1954
Hydrozoans					
Hydra	veg	18	75	no	Turner 1950
Turbellarians					
Stenostomum	veg	17	200	?yes	Hartmann 1920
	veg	150	120	yes/no	Sonneborn 1930
Phagocata	veg	13	1000	no	Child 1914
Dugesia	veg	30–40	?	no	Lange 19
Rotifers					
Philodina	egg	50	900	no	Meadow and Barrows 1971
Epiphanes	egg	503	1000	yes	Whitney 1912
Euchlanis	egg	17		no	Lynch and Smith 1931
Oligochaetes					
Aelosoma	veg	26	170	no	Maupas 1919
	veg	175	300	yes/no	Hammerling 1924
Aulophorus	veg	150	800	no	Maupas 1919
Chaetogaster	veg	43	140	no	Maupas 1919
Nais	veg	52	500	no	Maupas 1919
Pristina	veg	45	500	no	Maupas 1919
Crustaceans					
Daphnia	egg	570	1000+	no	Banta and Wood 1937

commonly reproduces by budding a succession of daughters from a parental frond which eventually ages and dies, a process described in more detail in the section below on the Lansing effect. According to Ashby and Wangermann the clone itself does not deteriorate, but they give no quantitative details.

7.2 Apomictic metazoans

The only asexual metazoans which have been reared in isolate culture for hundreds of generations are the rotifer *Epiphanes* (*Hydatina*) and the crustacean *Daphnia*. Both are apomicts, with the strictly mitotic production of asexual eggs, but can also reproduce sexually.

Whitney (1912) took *Epiphanes* through up to 503 generations in $2\frac{1}{2}$ years. Unfortunately, he did not carry out any regular census of his lines, but after 400 generations mortality and sterility were much increased, and the fecundity of the surviving fertile females had fallen. I have summarized his data as best I can in Figure 30; this is not very satisfactory, but it remains the only picture we have of long-term clonal senescence in a metazoan. Whitney also studied the effects of sexuality. This was shortly after the rediscovery of Mendelism, when Castle, Shull and others were

Figure 30. Clonal senescence in *Epiphanes* (rotifer). Letters refer to different clones, though B is an asexual derivative of A. Source: Whitney (1912), Tables 1–9.

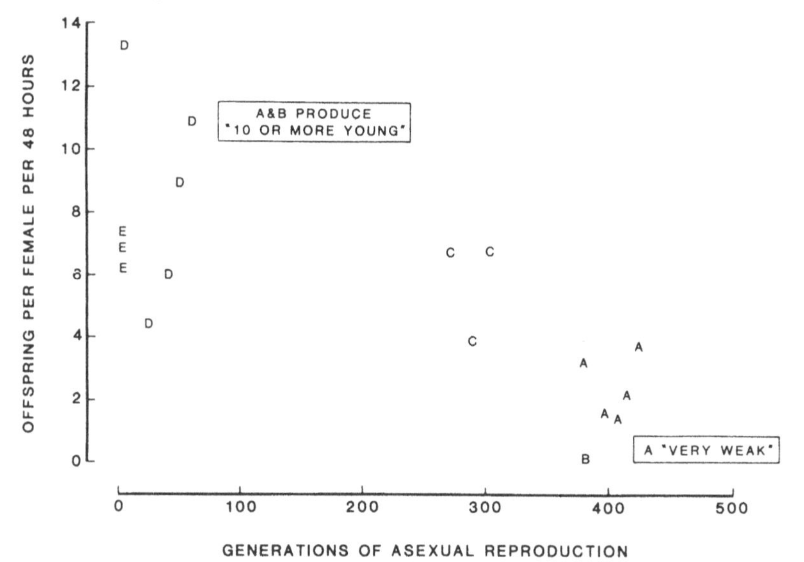

developing modern concepts of heterosis and inbreeding depression. Consequently, Whitney's experiments are not well-designed, and his descriptions of them are in places obscure. They are summarized in Figure 31, which gives the result of each mating in terms of the mean fecundity per unit time of individuals hatching from fertilized eggs; it is not possible to calculate variances from his data. The line involved was initiated with a single fertilized egg. After sixty generations it was split into two sublines, which were propagated separately for a further 300–350 asexual generations. Whitney then performed successive self-fertilizations within

Figure 31. Summary of Whitney's experiment with *Epiphanes* (rotifer). A_x is the x-th successive clonal self-fertilization within line A; A_xB_y is the cross between individuals produced by x successive clonal self-fertilizations in line A and y successive clonal selfings in line B. Numbers are fecundities per female per 48 h; the two values for A_2B_3 represent the two reciprocal crosses. Source: Whitney (1912), Tables 1–9.

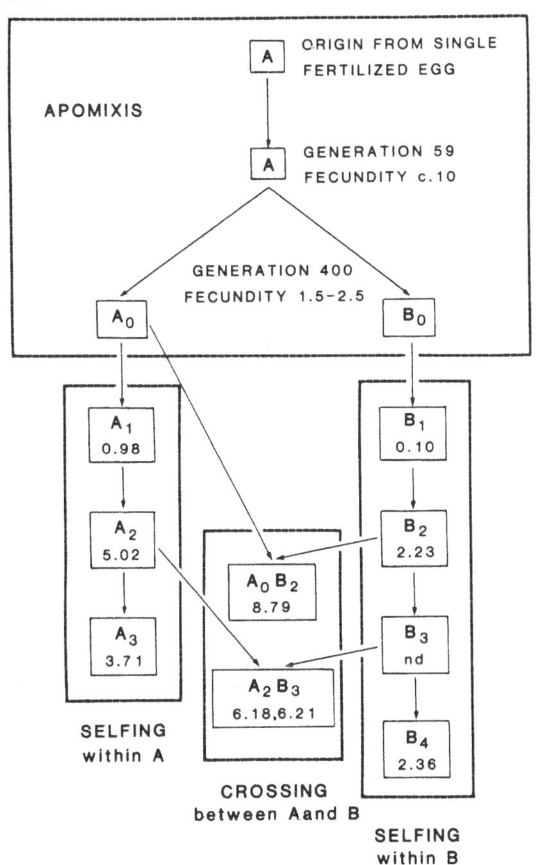

each subline, and made crosses between them. The first generation of clonal selfing had very low fecundity, which recovered in subsequent selfings, so as to exceed the parental line. Crosses between the sublines produced still greater fecundities, approaching those of young clones isolated from wild material. Whitney interprets his data as showing that inbreeding has little or no effect in raising the vitality of an old clone, while outcrossing is capable of rejuvenation. However, the only real difference between his selfed and crossed matings is the length of asexual descent of the partners from a common ancestor. This amounts to a few generations for the self-fertilizations and a few hundred generations for the crosses. The deterioration of the lines during strictly asexual reproduction is presumably due to the accumulation of partly recessive deleterious mutations; the reduction in mean fitness on selfing represents the segregation of these mutations; and the partial recovery in subsequent selfed generations reflects the prior elimination of recessive homozygotes. The effect is not as striking as in ciliates; but, unknown to Whitney, male rotifers are haploid. Consequently, any mutations which are severely deleterious when exposed in the hemizygous state in males will not participate in the matings, and cannot be made homozygous in the daughters. The enhanced performance apparently associated with crossing may indicate some degree of heterosis, but the crosses are not sufficiently well-defined for one to be sure. There is no difference between the two reciprocal crosses made, and therefore no evidence for any cytoplasmic degeneration. In short, Whitney's observations are consistent with the ciliate results, in that germ-line senescence during asexual reproduction is caused by the accumulation of mutations in nuclear genes, and can be at least partly reversed by meiosis and fertilization.

Jollos (1921) claimed that Whitney's results were caused by the cumulative effect of poor conditions of culture, and has some data to back up his argument. However, this could not be confirmed by Lynch and Smith (1931). They found a decline in three measures of vitality over only 15 generations in well-fed cultures of the rotifer *Proales*, as shown in Figure 32. A similar decline occurred in cultures maintained at low rations, but restoring an adequate food supply immediately restored vigour to the level of well-fed control cultures of the same clonal age.

Banta and Wood (1937; see also Banta 1914 and 1915) give a very brief account of a very long-lived culture of *Daphnia longispina*. They do not mention any senescent change after 600 asexual generations (the clone was still 'vigorous'), but the mean viability and hatchability of individuals produced by clonal selfing declined as the clone aged. At the same time, the variance of a number of reproductive characters was markedly greater

Figure 32. Decline in performance of the rotifer *Proales* over 15 asexual generations. The three independent characters plotted are: hatchability, the proportion of eggs hatching; longevity, the mean lifespan of individuals which did hatch; and fecundity, the rate of production of eggs as mean total number of eggs produced per female which hatched, divided by mean longevity. Each is plotted as a fraction of its value in the first generation. Two series are shown, symbolized by open and solid circles. The regressions are shown by solid lines and the slopes given; these should not be taken too literally (the data were not transformed, and the two series are in any case heterogeneous), but indicate the negative trend in the data, and the difference in the magnitude of this effect between the three characters. Source: Lynch and Smith (1931), the full-strength series.

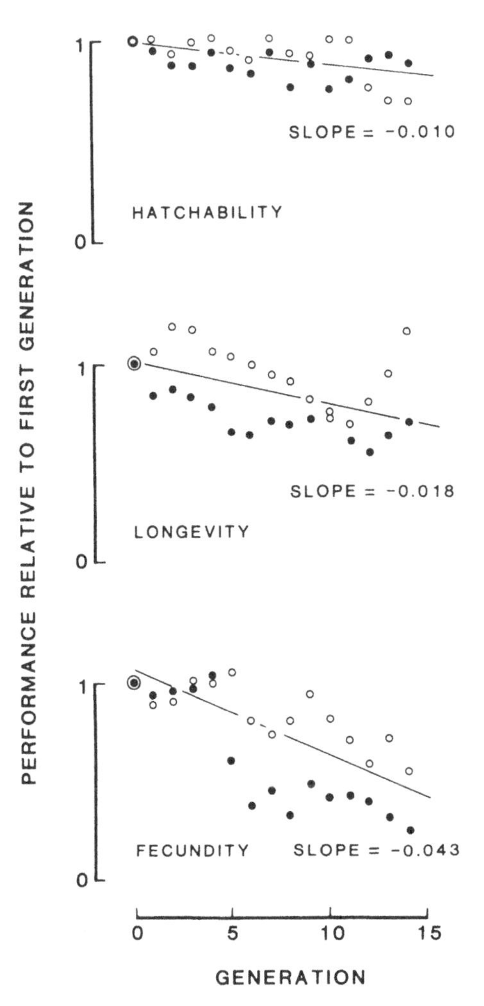

in sexual than in asexual families. They attribute these observations, quite correctly in my view, to the accumulation of deleterious mutations, in this case almost completely recessive, through time.

7.3 Vegetative metazoans

In apomicts like rotifers and cladocerans each line of descent is reduced at some point in every generation to the dimensions of a simple protist – a single cell bearing a single copy of the genome. Some metazoans, however, reproduce vegetatively by budding or fission. In this case, the line of descent is never reduced to smaller dimensions than the group of cells which is proliferated from the parent and undergoes morphogenesis to form the offspring. In very many and perhaps all cases of vegetative reproduction the formation of the new individual is connected with a special category of undifferentiated stem cells, which are also involved in tissue replacement and regeneration. They have been particularly well studied in hydras and planarians. Stem cells are akin to germ cells in that they maintain the continuity of a line of descent from generation to generation, and several authors (especially Brien 1953) have suggested a closer parallel by claiming that stem cells possess a Weismannian immortality, and need never wear out or decay.

Hydras can be maintained for years in laboratory culture, budding frequently and showing no signs of senescence. The animal is in a state of continual growth, so that if any region of the epithelium is marked, the mark will subsequently move up towards the hypostome or down towards the pedal disc. Specialized cell-types, the nematocysts and gland cells, are replaced from a self-replicating population of stem cells, the interstitial or I-cells, which are also involved in wound regeneration and bud formation. They do not alone contribute to such processes, since when the entire I-cell population is destroyed with nitrogen mustard, regeneration and budding still occur; moreover, complete small hydras can be regenerated from explants consisting entirely of epitheliomuscular tissue (see reviews in Burnet 1973), so that it is certain that dedifferentiation of specialized cells can occur under some conditions. Nevertheless, hydras without I-cells do not live long, and cannot form gametes. The concept of a population of asexual stem cells involved in vegetative reproduction therefore retains a great deal of force.

The distinction between individual and clonal senescence is not always sharply drawn in studies of hydras. Brien (1953) is quite clear that he means both: 'Non seulement la lignée d'Hydres se perpetue indefiniment mais l'individu Hydre est immortel' (p. 341). The vigour of this statement is however, more impressive than the quality of the data on which it is based. Individual polyps have been kept alive for more than two years (Goetsch 1922, 1925; Brien 1953), though I have not found any reliable records extending as far as three years. This has led some workers to claim that individuals are potentially immortal, while others have found a limited life-span culminating in depression, dedifferentiation (tentacle loss, for example) and dissolution. The depression of hydras, however, which is familiar to anyone who has cultured them, is quite different from the depression of ciliates, in that it is readily reversible, so that a vigorously budding individual can regenerate from what was to all appearances a decaying stump. Indeed, Reisa (1973) has interpreted depression as a reduction stage which carries the polyp through a period of unfavourable conditions by reducing metabolism and maintenance costs. The most satisfactory way of detecting senescence is to plot the rate of survival or reproduction on age. The only long-term data on the budding rates of individual polyps was published by Turner (1950). There is an increase rather than a decrease up to about 50 days of age; beyond that age fecundity seems to level off or decline, but the data are too sparse to say for sure (Figure 33(*a*)). My own observations, an example of which is given in Figure 33(*b*), show rather clearly an increase in budding rate up to about 20–25 days old, and thereafter a decline. Prof. L. Slobodkin, however, assures me that if the hydras are maintained properly there is no decline in fecundity with age. The survival rate of polyps is given by Pearl and Miner (1935), who tabulate data obtained by Hase (1909). This has ever since been cited as the classical example of a 'diagonal' survival curve, with the rate of survival independent of age. It is in fact the classical example of the inutility of plotting the number surviving to age x, $N(x)$, on age x; successive values of $N(x)$ are so strongly autocorrelated that such plots are likely to mislead, as well as being unsuitable for any conventional statistical analysis. The correct procedure is to plot the survival rate from age x to age $x+1$, $N(x+1)/N(x)$, on age x. If the rate of survival is independent of age, then the regression of $N(x+1)/N(x)$ on x has a slope of zero and an intercept equal to the mean rate of survival. If the slope of the graph is negative, a senescent decline in the rate of survival with age has been proven. In fact, Hase's data as abstracted by Pearl and Miner (I have not seen the original paper), quite clearly shows a gradual decline in the rate of survival from about 0.956 per five days in very young animals to 0.897 at age 90 days, after which it declines precipitously to about 0.5 by age 135 days. A glance at Pearl and Miner's table, however, shows that these changes are of almost mathematical regularity. I presume that this extensive 'data' has been obtained from a smooth curve of some kind; certainly it is not possible to accept it as evidence for anything.

Turner (1950) also reported the mean budding rate of 18 successive vegetative generations (Figure 34). There is obviously no sign of a decline, but the series is too short to give a conclusive result.

Figure 33. Budding rate in *Hydra* as a function of age. (*a*) Turner (1950). Figures 3–5. 52 asexual individuals all derived from a single parent by up to 18 asexual generations. (*b*) personal observations. Clone of 24 individuals. In both cases, values plotted are mean fecundity for all surviving individuals at a given age.

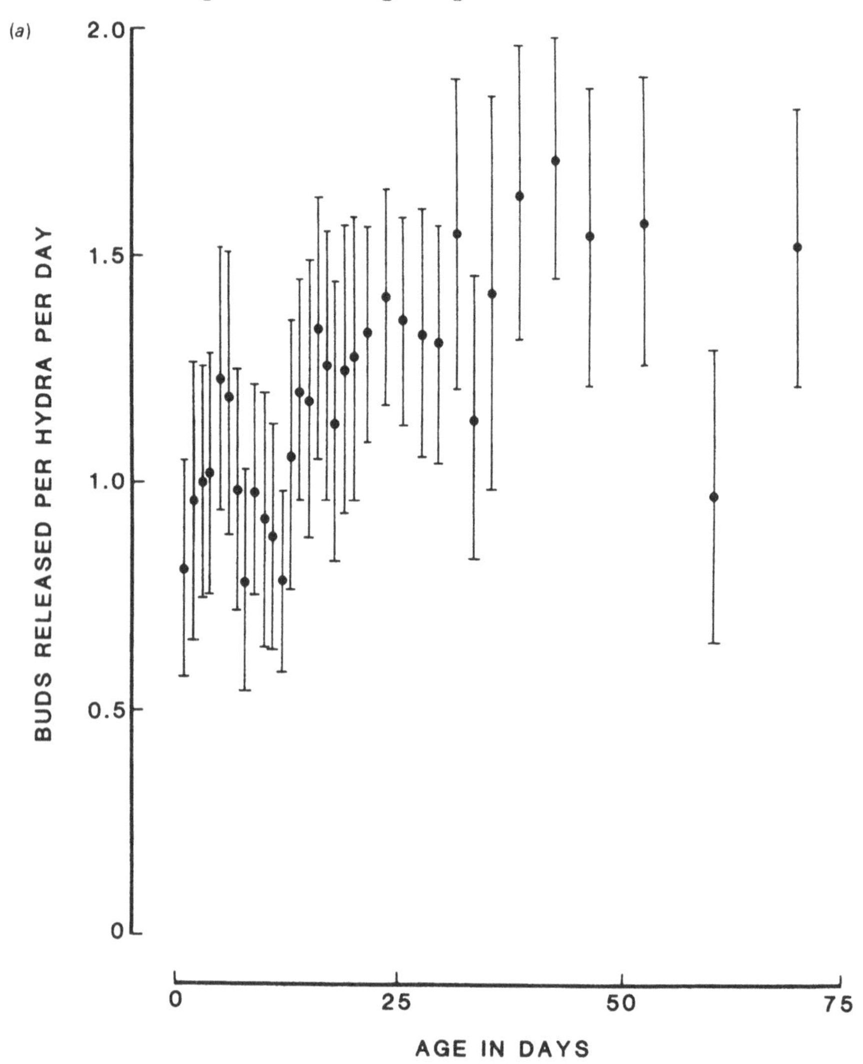

Like hydras, flatworms have a self-perpetuating population of stem cells which are involved in regeneration and vegetative reproduction, and when these are sterilized by irradiation the animals are unable to replace differentiated somatic tissue and die (Lange 1981). By repeated section of the body, an individual can be kept alive for many years with no apparent signs of senescence (Melander 1963). This is reminiscent of Hartmann's observation that individual *Amoeba proteus* can be kept alive for 130 days by repeated amputations of small pieces of the cell (Hartmann 1924), or the extension of the lifespan of the giant uni-nucleate alga *Acetabularia* by cutting off parts of the stalk (Hackett *et al.* 1963). Moreover, Child (1914, 1916) was able to maintain individual *Phagocata* (*Planaria*) for nearly four

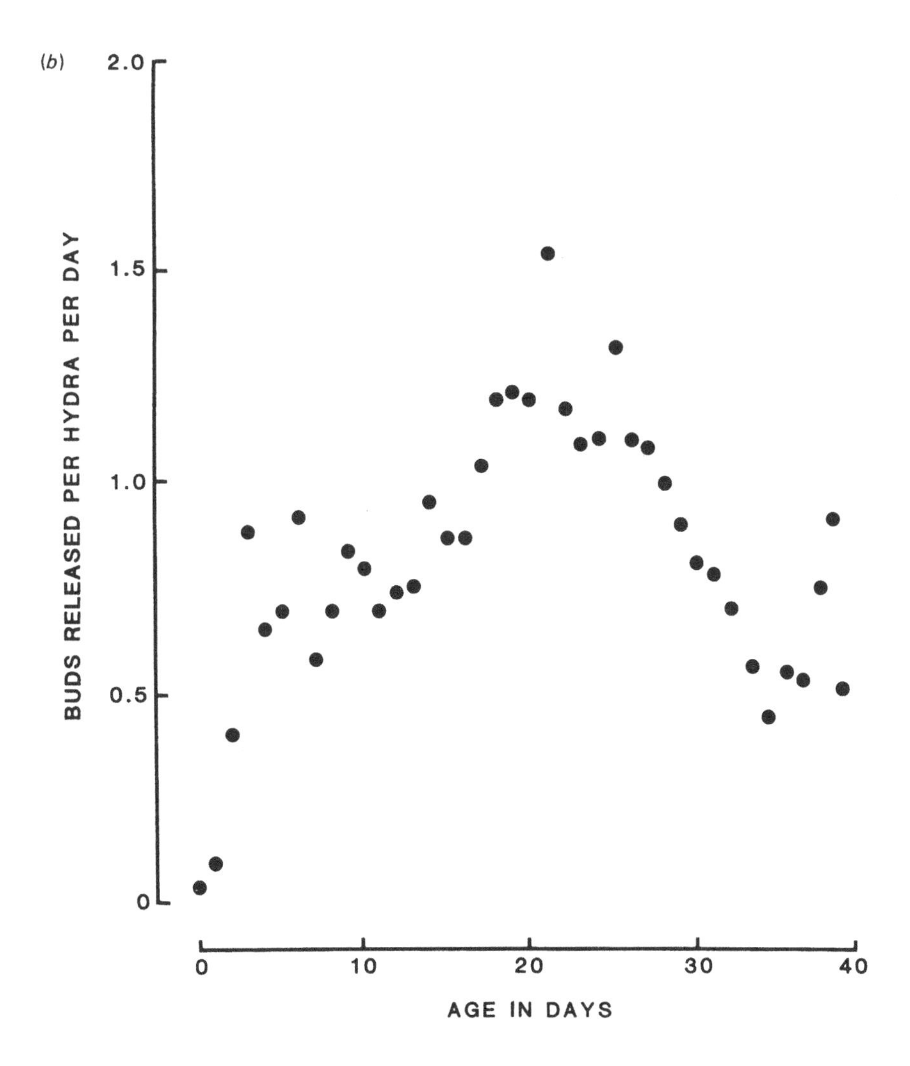

years by restricting their diet so that they neither increased nor decreased in size, just as Danielli and Muggleton (1959) were able to maintain amoebas on a ration that would not quite permit growth and division.

The apparent rejuvenation induced by cutting the worm into two has prompted speculations that lines perpetuated by the normal method of asexual reproduction in planariid turbellarians, a transverse binary fission, may be immortal. I can find no data on this point, but Child (1911) took *Phagocata* through 13 asexual generations without appreciable senescence. In these stocks, each worm fragments into a number of small pieces after a period of growth. Each fragment encysts individually, and after a period of differentiation a small worm emerges from the cyst to repeat the cycle. The 13 generations took about three years to complete.

Figure 34. Mean budding rate in 18 consecutive asexual generations of *Hydra*. Source: Turner (1950), Table 3.

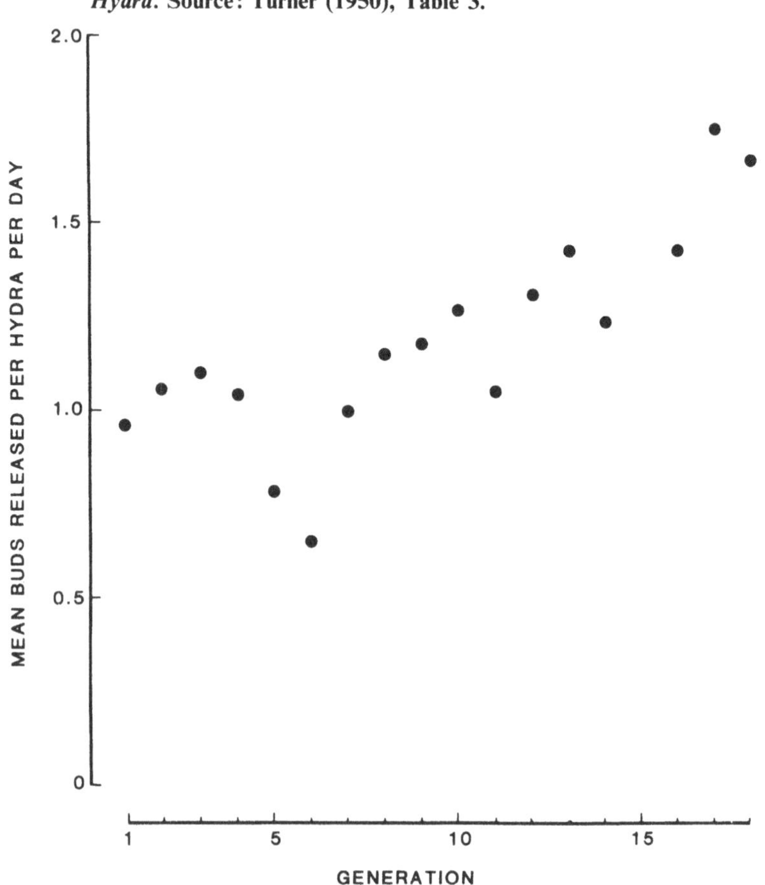

The classical study of clonal senescence in a fissiparous animal was performed by Sonneborn (1930) on another platyhelminth, the rhabdocoel *Stenostomum*. In these animals a fission zone appears about halfway down the body, and a new head develops at this site; the two fission products are thus differentiated to a large extent before separation, though the anterior fragment (the head end of the old worm, which is regenerating a new tail) and the posterior fragment (the tail end of the old worm, now regenerating a new head) are easily distinguishable for some time, since the posterior fragment is initially smaller and the regeneration of the head is not complete at fission. The difference in size means that the posterior end must grow before itself being capable of division; thus, the next fission takes about twice as long to occur in posterior as in anterior fragments,

Figure 35. Fission rate as a function of age in *Stenostomum*. Solid circles are anterior fragments (mean of 10 lines), open circles posterior fragments (mean of 4 lines). Source: Sonneborn (1930), Table 1.

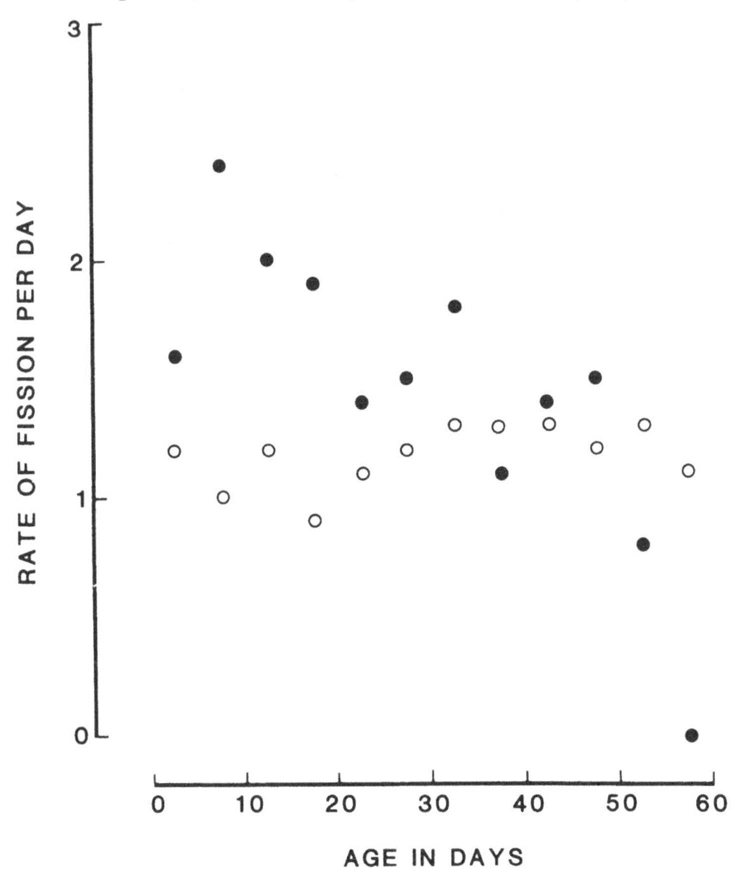

and if we were to set up two series, one from successive anterior and the other from successive posterior fragments, the anterior line would initially reproduce at twice the rate of the posterior line. However, as time goes on, the fission rate of the anterior line drops, eventually to zero, while that of the posterior line remains constant (Figure 35). Moreover, all of Sonneborn's 388 anterior lines died out, while 43/47 posterior lines were still alive when observations were terminated. While this is not conclusive (the posterior lines were not usually maintained for as long as the anterior lines), Sonneborn's data do show a fairly regular decline in the rate of survival with age in anterior fragments (Figure 36); no corresponding data for posterior fragments are available.

Sonneborn attributed the death of lines propagated from successive anterior fragments to structural degeneration associated with the appearance and enlargement of peculiar opaque bodies in the pseudocoel. These regularly appeared in anterior fragments, which regenerate only a short posterior region of the trunk, while retaining the old head and the

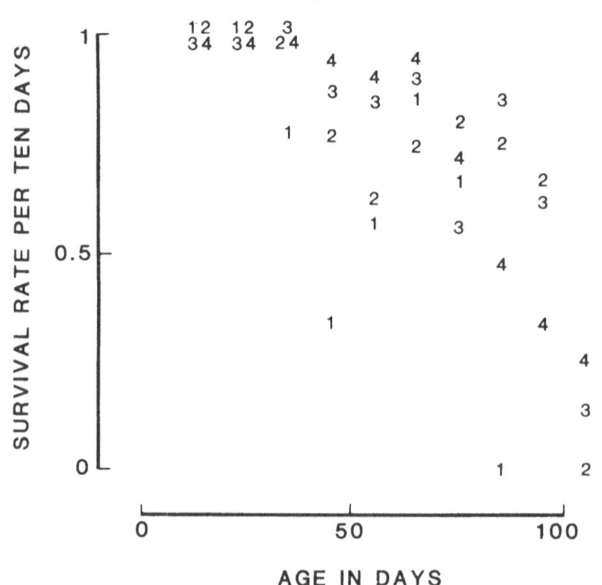

Figure 36. Survival rate as a function of age in anterior fragments of *Stenostomum*. Numbers refer to different lines, lines 2–4 being derived as posterior fragments from line 1 at ages 10, 20 and 30 days respectively. Source: Sonneborn (1930), Table 5. No sample size is given, but from internal evidence it is probably 30–50 worms per line. This number declines with age because of mortality, so that standard errors will be very large for ages in excess of about 80 days.

pharyngeal region usually affected by these deformities; they are rarely observed in posterior lines, in which the head and most of the trunk are newly formed. This led Sonneborn to interpret that anterior fragment as the parental individual, which senesces and dies like the adults of animals which reproduce by eggs, and the posterior fragment as the offspring. He went on to show that the lifespan of an anterior line could be prolonged by rederiving it at intervals from a posterior fragment. In this way, the lines were maintained for 320 days, about three times the maximum lifespan of purely anterior lines, without apparent degeneration. However, in the latter stages of degeneration the bodies in the pseudocoel extend across the fission zone, and if the line is then rederived from a posterior fragment showing signs of structural abnormality it quickly dies out. Sonneborn's study emphasizes the difficulty of distinguishing between the senescence of an individual and the senescence of a line in vegetatively reproducing organisms. It should also be borne in mind that, if the decay of his anterior lines is the classic instance of the senescence of a vegetative line, it is also to the best of my knowledge the only such instance.

It is likely that similar phenomena occur in small aquatic oligochaetes in the families Aelosomatidae and Naiadidae, though the data are not straightforward. Maupas himself cultured several species, and found that they maintained more or less constant rates of fission (Maupas 1919). He did not distinguish between anterior and posterior fragments – fission is more nearly equal in these worms than in *Stenostomum* – and his results cannot be assessed quantitatively because of wide variations in temperature. Hammerling's work on *Aelosoma* (Hammerling 1924) has often been cited as another instance in which anterior fragments are mortal while posterior fragments are not, but I find no good evidence for this in the original paper. The fission rates for two long-lived series are shown in Figure 37(*a*). There is no obvious decline with age in either the anterior or the posterior line; the only difference between the two is that the anterior line seems much more variable, but this may also be attributable to variation in temperature. Hammerling does not give any useful information on survival.

I have myself propagated anterior lines of *Aelosoma* and *Pristina* in isolate culture (Bell 1985a). Like Sonneborn and Hammerling, I observed that death often followed a progressive diminution in size, sometimes culminating in extensive dedifferentiation and dissolution, or pathological changes associated with the growth of structures in the body cavity. However, the incidence of these causes of mortality seemed to be independent of age: there was no indication whatsoever that survival rate decreased in older animals. Fission rate was low for the first few days after

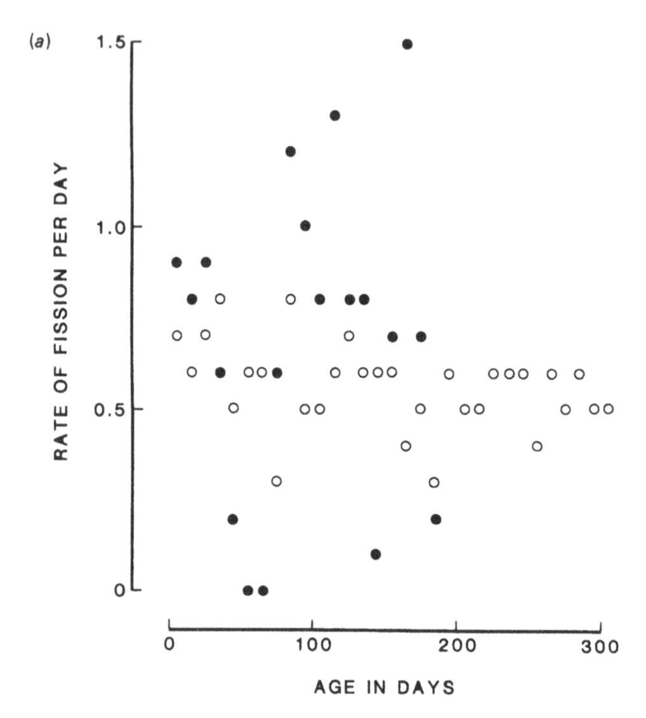
(a)
1.5
1.0
0.5
0
RATE OF FISSION PER DAY
0
100
200
300
AGE IN DAYS

(b)
1.0
0.9
0.8
0.7
SURVIVAL RATE

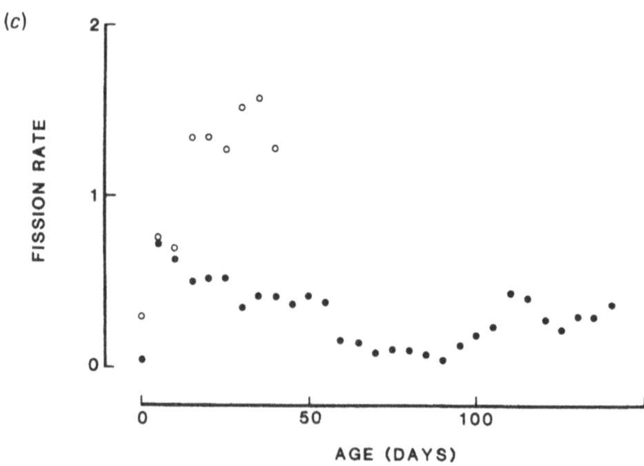
(c)
2
1
0
FISSION RATE
0
50
100
AGE (DAYS)

separation, then rose to a maximum, and in older animals fluctuated around a somewhat lower value. My observations on *Aelosoma* are summarized in Figures 37(*b*) and (*c*); the data for *Pristina* are similar. They contrast sharply with observations made at the same time on oviparous animals such as ostracods, cladocerans and rotifers, all of which followed a well-defined schedule of senescent decline.

7.4 The Lansing Effect

The analogy between the more or less rapid and inevitable aging of the soma and the much slower and more cryptic deterioration of the germ line has led several biologists to whether the two might not be directly related. It is, of course, a familiar observation that the offspring of older mothers are often less viable or fertile; in some organisms, such as rotifers, this effect is very pronounced (Jennings and Lynch 1928). However, there is no difficulty in imagining ways in which an aging soma might be unable to produce uninjured or adequately provisioned eggs. The crucial experiment is to isolate the first offspring of a founding asexual female, and in turn to isolate the first offspring of these individuals, and so forth, to construct a line in which each individual has descended through many generations from offspring produced early in life. This line is known as an 'orthoclone', in this case an early orthoclone, or pediaclone. In any given generation, its survival and fecundity schedule can be compared with the corresponding orthoclone derived from late-born offspring, a geriaclone. The classical experiment of this sort was performed with the bdelloid rotifer *Philodina* by Lansing (1947, 1954). He found that geriaclones cannot be perpetuated for very long, because of a cumulative decrease in fecundity and egg viability; this is known as the

Figure 37. Survival and fecundity schedules in *Aelosoma*. (*a*) Fission rate in anterior (solid circles) and posterior (open circles) lines. These two lines illustrated are the two longest-lived reported by Hammerling (1924), Table 11 and 13. (*b*) Survival rate of two lines of *Aelosoma*: open circles are a line cultured on rich (alga) medium, solid circles a line cultured on poor (bacterial) medium. Personal observations; see Bell (1985a). (*c*) Fission rate in anterior lines, using the same symbolism. Personal observations; see Bell (1985a). My cultures are described more fully in Bell (1984, 1985a), where it is shown that there is no trend in survival rate with age either in *Aelosoma* or *Pristina*. The fission rate data are too erratic to interpret straightforwardly; fission rate in the longer-lived culture first increased, then gradually fell, but at last recovered again. The two lines analysed comprised 115 (open circles) and 211 (solid circles) worms.

Lansing effect. The effect is readily reversible; the vitality of deteriorating geriaclones is restored in one generation by choosing early rather than late offspring to found a new line. These very striking observations aroused a great deal of interest, and several authors attempted to reproduce Lansing's result. The most extensive and careful work is King's study of the monogonont rotifer *Euchlanis* (King 1967). He does not describe successive generations, but a census after 10 or more generations showed conclusively that the geriaclone had a lower birth-rate and a higher death-rate than the pediaclone, when cultured under any of a variety of conditions. Indeed, in practice the geriaclone could not be perpetuated, and twice had to be refounded from early offspring.

Unfortunately, a second attempt to confirm Lansing's result failed to do so. Meadow and Barrows (1971), who like Lansing used *Philodina*, found no difference in lifespan or fecundity between pediaclones (cultured for 50 generations) and geriaclones (cultured for 16 generations), whether maintained with one species of food or two. What they did find was that the entire fecundity schedule was shifted towards later ages in the geriaclones. In a sexual organism there would be no difficulty in interpreting this result: by choosing eggs produced early or late in life by a group of females one is automatically selecting for early and late reproduction respectively. The same effect might occur in an asexual organism, if the base population were heterogeneous. But *Philodina* was chosen for these experiments precisely because bdelloids are thought to be exclusively asexual, and Meadow and Barrows claim to have cloned their material to ensure genetic uniformity.

Several experiments using sexual organisms such as *Drosophila* have also been reported (e.g. Comfort 1953, Wattiaux 1968). I think it is fair to say that these have succeeded only in adding to the confusion which surrounds both the validity and the interpretation of the Lansing effect. For example, both Lints and Hoste (1974), working with *Drosophila*, and Beguet (1972), using the autogamous nematode *Caenorhabditis*, found that the vitality of the geriaclone declined for several successive generations, but then returned to about its original level. The only obvious conclusion is that sexual organisms, in which any Lansing effect may be either swamped or mimicked by variation and selection, are inappropriate for this sort of work.

Finally, there are certain curious organisms in which there can be no doubt of a cumulative but reversible effect of maternal age on offspring quality, though not in the way that Lansing envisaged. These are the few organisms which produce offspring larger than themselves. The two examples I know of are diatoms and duckweed. The diatom frustule is shaped like a box, with a lid and a bottom piece. During asexual fission the two halves separate, and each forms the lid of a daughter cell; thus, one daughter becomes as large as its parent, while the other is smaller. In asexual populations there is therefore a continual decrease in mean size, which can be arrested only when two small cells fuse to form a sexual auxospore. In duckweed (*Lemna*), an individual frond produces a succession of about a dozen daughter fronds over a lifespan of 40–50 days. The daughter fronds produced early in life are as large as the parental frond, but later daughters are much smaller (Ashby *et al.* 1948). A more or less constant mean size is preserved because the first grand-daughters produced by a small, late daughter are relatively large; the effect is cumulative, and it may take up to six generations to restore the frond area of the original parent. In both diatoms and duckweed, then, there is a cumulative and transmissible decrease in size, which may affect viability and fecundity. The duckweed case is the clearer, and indeed Ashby states explicitly that the daughters produced by aging mothers have shorter lives and fewer offspring than their sisters, produced when their mother was youthful.

The Lansing effect has so far produced little beyond a vague sense of unease. It may amount to nothing more than the trivial outcome of well-known phenomena, or it may tell us something interesting about the aging of germ cells. The confused and contradictory experimental results have tended to inhibit theoretical work, and it is not likely that this situation will change until some plausible mechanism is clearly described and used as the basis for much sharper experimental designs.

7.5 Metazoan cells in tissue culture

The high period of protozoan isolate lines was also the time when the technique of tissue culture was developed. From the first it seems to have been anticipated that isolated cells might be immortal, and this was confirmed by Carrell (1912) and Ebeling (1913), who maintained chick embryo fibroblasts for up to 34 years before their cultures were voluntarily terminated. After similar results had been obtained with mouse mesenchyme (Earle 1943) and human tumour cells (Gey and Gey 1936), the indefinite lifespan of somatic cells outside the body seemed firmly established. This was a fundamental result, since it implied that senescence and death must somehow arise from the relationships between differentiated cells, rather than being an intrinsic property of the cells themselves. Despite the weight of evidence in its favour, however, it was overturned decisively in a series of papers by Hayflick (Hayflick and Moorhead 1961;

Hayflick 1965), who showed that fibroblast-like cells from various human tissues maintained a constant rate of multiplication for only a finite period of time, after which they rapidly aged, with a decrease in vigour ending in the extinction of the line. This is now universally accepted as a correct generalization, though one to which there are several categories of exceptions.

In Hayflick's cultures, cells derived from foetal human tissue divided about 50 (± 10) times before senescence set in. This is a remarkable figure. A metazoan cell has a typical dimension of 10^{-4} m and therefore a typical mass of about 10^{-9} kg. Since the mass of the whole body is about 10^{2} kg, it must contain about 10^{11} cells, which would arise from the zygote by $11/(\log_{10} 2) = 37$ successive divisions. This suggests irresistibly that foetal cells have the innate capacity to divide just often enough for the individual to complete development, after which they begin to senesce. I do not know how close this arithmetic coincidence really is – the foetuses from which the tissue is taken are already three months old, but on the other hand some tissues, such as blood and epithelium, are continually renewed – but the idea that the onset of senescence in cultured cells is somehow tuned to the timescale of the body from which they came has been supported by several lines of evidence. In the first place, it is the number of divisions, rather than chronological age, which seems to determine the onset of senescence, since experiments in which cell division is arrested, for example by nutrient restriction, usually (Dell'Orco *et al.* 1974, Goldstein and Singal 1974) though not invariably (McHale *et al.* 1971) show that the length of time spent in the arrested state has little effect on the time of onset of senescence. Secondly, the number of divisions achieved in culture appears to be a declining function of the age of the donor (Martin *et al.* 1970, 1981; Schneider and Mitsui 1976). The regression is rather shallow, however, so that tissue from 70-year-old donors can still achieve about 30 divisions, and there is a good deal of scatter around the trend, with r^2 only 0.1–0.2 or so. Finally, there is some evidence that cells from short-lived species senesce earlier than those of long-lived species. Very long-lived cultures, for instance, were obtained from giant tortoises. Sacher and Hart (1971) compared *Peromyscus* with *Mus* since, while comparable in many other respects, the former lives 2–3 times as long; fibroblast cultures from *Peromyscus* were correspondingly 2–3 times longer-lived than those from *Mus*. However, there has been no careful comparative work across a wide range of species.

The relevance of these observations to somatic aging and death in intact metazoans remains in some doubt (see Kohn 1981). Indeed, virtually the only direct evidence of a link between the finite proliferation of cultured cells and the finite lifespan of individuals is the rapid senescence of cultures derived from patients with one of the diseases associated with premature senility (Goldstein 1969, Danes 1971, Menhaus *et al.* 1971). On the other hand, it must be admitted that cell loss, or even lower division rate, does not seem to contribute substantially to normal mortality. Indeed, death is much more likely to be caused by the unrestrained growth of neoplastic tissue than by any progressive reduction of cell number.

Moreover, some cell cultures are, if not immortal, at least extremely long-lived. Many rodent tissues, in particular, continue to proliferate for 500 divisions or more, or least an order of magnitude in excess of cultured human cells (Cameron 1972, Petursson *et al.* 1964, Krooth *et al.* 1964). It is also likely that tissues involved in regeneration or early embryogenesis, such as human haematopoietic cells (Moore and Sandberg 1964), lizard tail regenerate (Simpson and Cox 1972) and *Drosophila* imaginal disc tissue (Hadorn 1967), senesce very slowly if at all. It would be of some interest to culture the stem cells of coelenterates, platyhelminths and annelids; so far as I know, this has not been done.

The most conspicuous exceptions to the general rule of finite longevity in culture are provided by tumour cells, which in many cases can be cultured or transplanted indefinitely. Such cells almost invariably have an abnormal karyotype. The precise chromosome complement varies within and between different types of tumour, but there is typically extensive aneuploidy around a mode suggesting tetraploidy. Unfortunately, the association between tetraploidy and 'immortality' on the one hand, and euploidy and senescence on the other, has led to a rather sterile debate in which different authors define karyotypic normality to suit their own hypotheses about cellular aging. It is abundantly clear that there is some link between karyotype and clonal lifespan, but the mechanism involved is entirely obscure.

To compare tissue cultures with protozoan cultures, it is necessary to know how tissue cultures are propagated. They are in all cases mass cultures: a cell colony is allowed to grow until it has doubled in size, at which point half the cells are transferred to a new plate. The number of cells involved in serial transfer is typically 10^6–10^7, and the technique is therefore comparable to the mass culture of ciliates in batch volumes of about 10^{-1} l. In drawing parallels between the senescence of cultured cells and protozoans, it must always be borne in mind that, while asexual isolate lines of protozoans usually or always senesce, mass cultures can normally be maintained for as long as the investigator wishes.

The interpretation of cell cultures hinges on the pattern of senescence. The decline in the average rate of division might represent a general senescence affecting all cells more or less equally, or it might represent the accumulation of cells incapable of dividing. This distinction will be made by isolating a set of single cells from culture and measuring the longevity of the mass cultures they give rise to (e.g. Smith and Hayflick 1974); alternatively, individual cell lineages could be traced within mass cultures by a technique such as time-lapse microcinematography (Absher and Absher 1976). All such studies seem to show substantial variance in clonal lifespan within the group of cells isolated at any particular time. This variation is not unlimited: the future lifespan of the *best* clone present at any given time seems to be about the same as that of the mass culture from which it was derived. This suggests that the process of tissue-culture aging is in great part the sequential replacement of cells with lesser by cells with greater division potential, so that the lifespan of the culture is determined by the lifespan of its longest-lived constituent.

Given such variation between cells, we would like to know how it arises and whether it is transmitted from mother to daughter. The variable of interest is the time spent between divisions. This seems to depend on cell size rather than on cell age (since the last division) *per se*, since age at division is about twice as variable as size at division. Naturally, this is consistent with the observation that repeated amputation of cytoplasm from unicells such as *Amoeba* and *Acetabularia* will postpone division indefinitely. Now, if the probability of entering division is an increasing function of cell size, it is easy to show that the interdivision periods of mothers and daughters will be correlated. A good review of the basic stochastic theory is given by Tyson (1985). Briefly, suppose that a given cell happens to divide after a relatively short time, while it is still small. Since its daughters will be small, they will probably spend longer than average growing before they themselves divide. Conversely, a parental cell which divides when it is large will produce large daughters, which can themselves divide at a relatively early age. The interdivision times of mothers and daughters will therefore be negatively correlated; in the model proposed by Tyson, which is very successful in predicting the stationary distributions of ages and sizes at division in yeast populations, the mother–daughter correlation is -0.5. The same argument leads to the conclusion that the correlation between daughters will be positive, being $+0.5$ in Tyson's model. In practice, these correlations may not hold in populations which do not have the unrestricted exponential growth assumed in simple models; in particular, with density-regulated logistic growth both the mother–daughter and the daughter–daughter correlations will move towards zero. Some observed correlations are shown in Figure 38. For bacterial and yeast cultures the daughter–daughter correlation is about +0.5, but the mother–daughter correlation, though generally negative, is only about −0.2. For cultured mammalian cells both correlations are much greater (positive) than stochastic size-dependent models seem to predict. One possible interpretation of these data is that the cultured cells, at least, possess some genetic variance for interdivision

Figure 38. The mother–daughter correlation r_{md} and the daughter–daughter correlation r_{ss} for various cultured cells. Solid circles are bacteria and yeast; open circles are cultured mammalian cells. The predicted correlations for exponential and logistic growth are also shown: the predicted value for exponential growth is $r_{md} = -0.5$, $r_{ss} = +0.5$; these should move in the direction shown by the arrow towards the hatched area close to 0, 0 when growth is regulated by cell density. Note that r_{md} and r_{ss} are themselves correlated (correlation coefficient = 0.51). Source: Tyson (1985), Table 2, plus four other cell culture systems calculated from figures in Absher and Absher (1974, 1976).

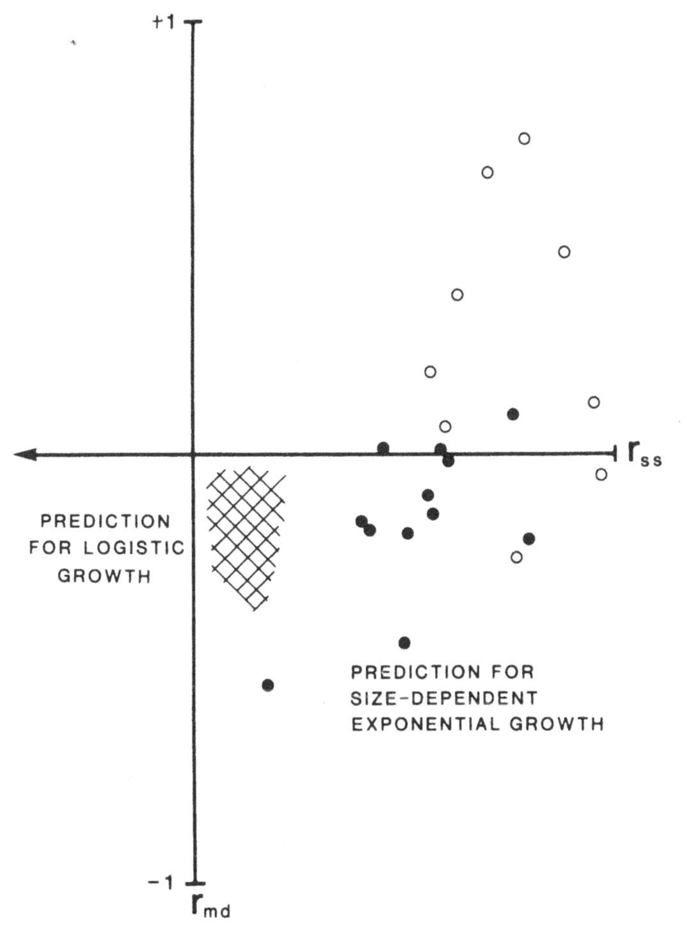

time. Positive correlations may also be generated, however, by any systematic change in division rate with time, and this seems the more likely explanation.

The source of any heritable variation between young and old cultured cells remains obscure. Wright and Hayflick (1975) enucleated one set of cells by cytocholasin treatment, and poisoned the cytoplasts of another set with iodoacetate and rotenone. They then used Sendai virus to fuse the enucleate cytoplasts with the nucleated cells whose cytoplasm was non-functional. They found that these chimaeras had the same lifespan as controls if the nucleus came from a young culture, regardless of whether the cytoplasm came from a young or an old culture; conversely, all the chimaeras died within about five divisions if the nucleus came from an old culture, regardless of the origin of the cytoplasm. This suggests that the determination of future lifespan is controlled by the nucleus, and not by the cytoplasm. However, the design of the experiment had some undesirable features, in particular the doubling of the volume of cytoplasm and the presence of a large quantity of poisoned cytoplasm. Muggleton-Harris and Hayflick (1976) attempted to avoid these objections by enucleating individual cells from cultures of different ages and then re-fusing nucleus and cytoplasm. Re-fusion of young nuclei into young cytoplasm produced cells whose lifespan was not grossly different from that of young controls. Fusing nucleus and cytoplasm of different ages produced a substantial decrease in longevity, but there appeared to be no difference between the young nucleus/old cytoplasm and old nucleus/young cytoplasm combinations. This seems to show that both nucleus and cytoplasm play a part in determining future clonal lifespan.

The precise nature of the parallel between the senescence of protozoan cultures and metazoan tissue cultures remains obscure. I suspect that it is a rather distant one. It is true that normal somatic cells usually senesce when cultured, and this may be due to the accumulation of harmful changes. However, the senescence of cultured cells is abrupt, rather than gradual. Nor are metazoan cells grown as isolate lines, but senesce in mass culture, which protozoans in general do not. Nor do I think that the senescence of isolated cells is likely to represent a major source of mortality for intact organisms. I think it is far more likely that cellular senescence is the byproduct of the senescence of the whole organism. Any gene which maintains a high level of cellular function, including fission, for the first few dozen divisions will be strongly favoured by selection, even if it causes a catastrophic loss of function thereafter. The reason for this is simply that the behaviour of the cell is immaterial if the body is shortly to die anyway. The senescence of the soma as a whole is not caused by the deterioration of individual cells; it follows from the decay and disintegration of tissues and organs, which will be favoured by selection provided that the genetic elements responsible also have the effect of enhancing early reproductive success. The general theory that life cycles evolve through selection acting on genes with different effects at different ages is described in more detail in the next chapter. It leads to the conclusion that the senescence of cultured cells is an outcome of the senescence of whole organisms, rather than vice versa. Cultures of protozoans and of metazoan tissues have little in common, and if both senesce, they do so for quite different reasons.

8

The Ratchet

8.1 Somatic assortment and clonal senescence

The only current theory of ciliate senescence which invokes a well-defined mechanism is the notion that it is the consequence of somatic assortment: if there is no regulation of the total number of copies of different elements, then continued amitosis will eventually lead to the production of macronuclei which completely lack some essential elements, as I have explained above in section 3.3. The onset of senescence then represents the appearance of these lethal nullisomics. This hypothesis is usually credited to Fauré-Fremiet (1953), but I have been unable to find a precise statement in his paper.

The mathematics of the process have been worked out by Kimura (1957). He showed that the probability P_t that a line will have lost at least one of n different types of element, each of which is initially present in m copies, by the t-th asexual generation, is approximated by

$$P_t \approx (1/\sqrt{2})\exp[(2\log 2 - 1)n - t(n-1)/4m].$$

This approximation was obtained for the case of large n, but works fairly well even for small n. For example, if the macronucleus of the founder contains 4 different elements, each present in 2 copies, Kimura's formula suggests that the median time to extinction (got by substituting $P_t = \frac{1}{2}$) is about 5.08 generations. By direct simulation, I obtained a mean value of 4.00 with a standard deviation of 1.56 generations, in ten replicates. Kimura points out that if the nucleus bears 100 copies of each of 41 elements (the values thought to be reasonable for *Paramecium* at the time) then 99% of lines will become extinct within 200 generations, which is consistent with observation. Ammermann (1971) later pointed out that the corresponding value for *Stylonichia*, then thought to possess about 60 copies of each of 150 elements, is about 100 generations; this is too short a period, but by further assuming that the macronuclear anlagen shed a great deal of genetic material, he was able to bring this value up to the 400 generations or so that *Stylonichia* usually lives for.

These numerical coincidences represent the main evidence for the somatic assortment hypothesis; so far as I know, there is no direct evidence for the assortment of nullisomics. Moreover, the coincidence between theory and observation may not be as close as the values I have cited above would suggest. Kimura's argument (and the simulations reported by Ammermann) refers to a *single* asexual line. In fact, almost all isolate cultures used four or five lines, with any line which goes extinct being restocked from a surviving line, should one be available. This will obviously increase the longevity of the series as a whole. I have simulated the effect of differing numbers of lines per series for a small macronucleus bearing 2 copies of each of 4 different elements (Figure 39). The mean time to extinction of the series t_E was related to the number of lines in the series L by the linear regression

$$\log_{10} t_E = 0.339L + 0.264,$$

suggesting that the addition of one more line to the series will roughly double the mean time to extinction. Now, for large n, Kimura's formula becomes very roughly, $t_E \approx 2m$, for the case of a series with a single line.

Figure 39. The longevity of an isolate series in relation to the number of lines maintained. These results were obtained from the simulation model described in section 5.1; cf. Figure 10.

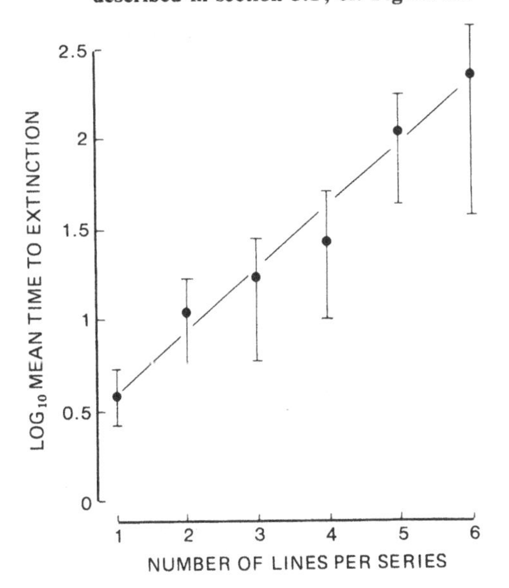

With L lines, therefore, we expect that $t_E \approx 2^L m$; since $L = 4$ or 5 in most cases, we have on average $t_E \approx 25m$.

This value is now rather too large, suggesting that isolate series should persist for 1000 generations or more. Recent work on macronuclear structure in hypotrichs such as *Stylonichia*, *Oxytricha* and *Euplotes* (reviewed by Raikov 1982 and Steinbruck 1986) puts it larger still. Even in hymenostomatids, the formation of the macronucleus involves the elimination of some parts of the micronuclear genome and the amplification of others. In *Tetrahymena*, the micronucleus contains about 0.21 pg DNA organized into 10 chromosomes, while the macronucleus contains about 30–50 times as much (Gibson and Martin 1971, Seyffert 1979). There are also qualitative differences between the two nuclei, about 10–20 % of micronuclear elements, probably moderately repetitive sequences, being absent from the macronucleus (Iwamura *et al.* 1979). The excision of these sequences implies that the macronuclear genome comprises a large number of relatively small elements: in place of the 10 chromosomes of the micronucleus, the macronucleus may contain about 30000 DNA molecules, about 650 kb long on average (Preer and Preer 1979). The break-up of each chromosome into about 70 fragments explains the lack of association during somatic assortment between elements which are known to be linked in the micronucleus. *Paramecium* is less well known, but here too the construction of the macronucleus appears to involve some degree of DNA elimination, with amplification of the remaining sequences (Schwartz and Meister 1975). Most macronuclear DNA molecules are rather large (greater than 20 kb, Steinbruck *et al.* 1981), but some may be much shorter, perhaps down to the size of individual genes (McTavish and Sommerville 1980). In *Glaucoma* the micronuclear chromosomes are broken up into short molecules 2–100 kb in length in the mature macronucleus.

These processes of genome fragmentation, elimination and amplification reach an extreme among hypotrichs. They have been described in detail for *Stylonichia* by Ammermann *et al.* (1974). After conjugation, the macronuclear anlage forms about 300 chromosomes, some 200 of which condense, move to the periphery of the anlage and are expelled into the cytoplasm. The remaining chromosomes despiralize and become polytenic, having much the same banded appearance as the salivary gland chromosomes of *Drosophila*. The bands of these giant chromosomes become marked off from one another by septa, and then separate completely, the chromosome falling apart. Over 90 % of these separate sequences are then eliminated from the anlage – again, it seems to be repetitive sequences that are thrown out (Steinbruck *et al.* 1981). The remaining sequences, representing only about 2% of the original micronuclear genome, then go through several successive rounds of replication to increase the DNA content of the mature macronucleus to about 1000 pg, as compared to the micronuclear value of about 25 pg. This DNA is now organized into about 10000 different, apparently gene-sized pieces, on average only about 3 kb in length, each of which has about 15000 copies (Prescott *et al.* 1973, Rae and Spear 1978). Other species differ in detail. For example, in *Oxytricha* all chromosomes become polytenic (Spear and Lauth 1976) and DNA elimination is generally less drastic, but the macronucleus still comes to contain about 2000 copies of each of about 20000 short sequences. The description of the macronucleus as 'a bag of free-floating genes' (Prescott *et al.* 1973), while perhaps oversimplified, seems nevertheless a good general summary of the situation in hypotrichs.

When the macronucleus contains several thousand copies of an element, the extinction of series by the assortment of nullisomics will require tens or hundreds of thousands of generations, two orders of magnitude in excess of observed values. Moreover, there should be a pronounced difference in longevity associated with the relatively simple hymenostomatid macronucleus as compared with the massively over-replicated macronucleus of hypotrichs. There is no sign of such a difference in the data, and hypotrichs such as *Stylonichia*, *Oxytricha* and *Uroleptus* provide classical examples of clonal senescence resulting in extinction within a few hundred generations.

There are certain other features of clonal senescence which resist interpretation in terms of somatic assortment.

First, although heterozygotes have often been shown to assort into homozygous lines, I can trace no observation of the 50% of 'nulls' that such assortment is expected to give rise to, if there is no control over copy number. Sonneborn *et al.* (1956) searched for such nulls, and for evidence of variance in gene dosage, without success.

Secondly, where adequate data are available the senescence of a series does not usually involve the sudden death or total sterility of an increasing proportion of its lines. Rather, there is a gradual decrease in vitality in all lines. We can accommodate this within the somatic assortment hypothesis by supposing that, while nullisomics are lethal, individuals which have relatively few copies of a given element are enfeebled. However, this introduces a serious complication, since selection can now act to eliminate unbalanced nuclei, and this tends to restore the complete complement of copies of a given element.

Thirdly, the hypothesis does not address senescent changes involving the micronuclear genome, such as the increased mortality of exconjugant lines observed when old clones are crossed.

Finally, somatic assortment cannot explain the senescence of clonal organisms other than ciliates, where nuclear division is mitotic.

Somatic assortment leading to the complete loss of essential genetic elements offers a simple and attractive interpretation of clonal senescence in ciliates. It may well be an important process, and deserves more searching direct tests than it has so far received. Nevertheless, the arguments I have set out above seem to show that it will not cover all the facts, and we must look elsewhere for a more comprehensive theory. So far as macronuclear degeneration is concerned, I think that such a theory is provided by the somatic assortment of deleterious mutations of individually small effect, whose irreversible accumulation will cause a gradual and progressive loss of function. This interpretation leads to much more general issues involving the integrity of the genome and the function of sex and recombination.

8.2 Muller's Ratchet

A very influential theory of sex, which dates back to August Weismann (1889), is that it provides the new combinations of genes necessary for adaptedness to increase under natural selection. In the simplest case, we might imagine a haploid organism which has lived in a particular environment for so long that, barring recurrent mutation, every locus is fixed for the allele best fitted to those conditions. The conditions then change, however, and at two loci genes which previously occurred only as rare mutations are now fitter than the wild-type alleles. These genes will spread through the population, of course, and will eventually become fixed. In an asexual population the process of fixation will be slow, since the only way in which the double mutation can arise is if the second mutation occurs in an individual which already bears the first. The newly-favoured genes must then be fixed successively: first one increases in frequency, so that the population comes to consist largely of this type, and only then will the double mutant be formed in appreciable numbers, enabling the population to reach its final state. Sex allows the population to evolve faster. The two mutations can arise in separate lines of descent, but through gametic fusion be recombined into the same zygote; the double mutant then appears at once, without having to wait for one of the single mutants to spread. This is the basis of Fisher's remark that even if an organism bears only two loci, sex will effectively double the rate of evolution; with more than two loci it will be even more effective.

I have reviewed this theory at greater length, and examined its merits as a general interpretation of sex, in a previous book (Bell 1982). It does not seem to contribute much to the rejuvenescence of protozoan clones. To extract a clone from a natural population and cultivate it in the laboratory is to expose it to a quite novel set of conditions, and it is perfectly plausible that conjugation will create many genotypes which are superior to their parents. Sex might in this way facilitate adaptation to novel conditions of culture. Indeed, the increased variance of exconjugants relative to conjugants (Figures 18 and 19) shows that this can happen: while the general effect of conjugation is to depress the mean rate of fission, a very few exconjugant lines exceed the range of the nonconjugants (Figure 18). Nevertheless, the effect of conjugation is not cumulative, in the sense of procuring a steady increase in performance (Figure 20). Rather, conjugation seems to restore previous levels of vitality. In the sections above, I have pointed out that it can do this by bringing together deleterious recessive genes so that they can be eliminated through the death of homozygotes. Sex, if you like, acts as an editor which detects serious copying errors and enables the genetic message to be transmitted without contamination. A general theory of this sort was first suggested by H. J. Muller (1964).

Imagine an infinite population of asexual haploid organisms. To begin with, every locus bears the gene which is best suited to the current environment. Even if the environment remains constant, however, the genome will not, because no copying process can work without error. There will be an inexorable increase in the number of deleterious mutations, which will slowly degrade the adaptedness of the population. This increase will be opposed by selection, which will always tend to remove inferior variants and restore the population to its original state. At some point, the two opposed processes of mutation and selection will come into balance. The population will then possess some frequency distribution of mutations per genome, with some lines having very few while others are heavily loaded. This distribution represents an equilibrium state which will be perpetuated indefinitely, so long as the environment does not change. Sex makes no difference to this process (at least in simple models where mutations at different loci have independent effects on fitness), and, if they are comparable in other respects, sexual and asexual populations will develop the same equilibrium frequency distribution of mutations per genome.

The situation is quite different in a *finite* asexual population. Supposing some frequency distribution to have developed, there will be an optimal category which bears the least number of mutations. This least-loaded line

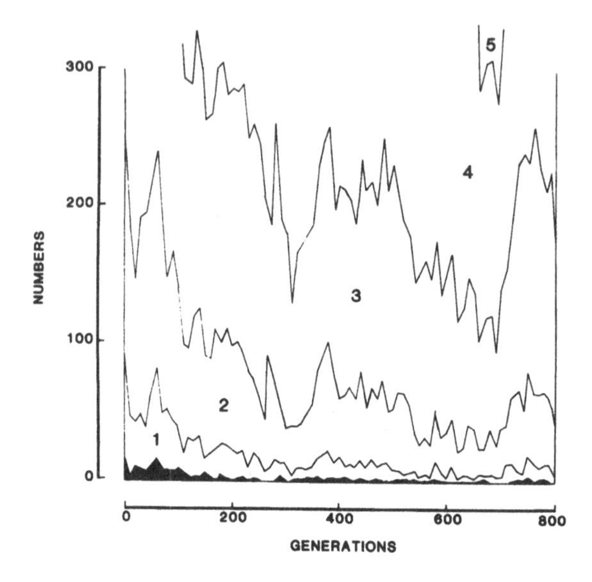

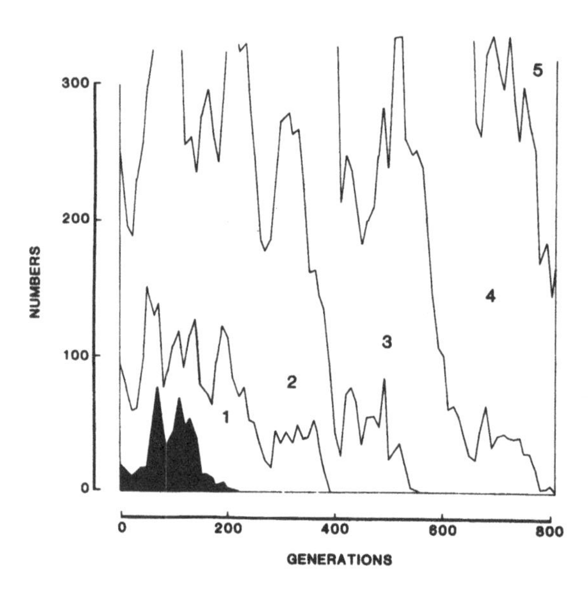

Figure 40. Muller's Ratchet. These diagrams show the output from Monte Carlo simulations of populations of 1000 individuals. Each diploid individual has a single chromosome bearing 5 diallelic loci, and the population is initially fixed for wild-type genes at all loci. In each generation, the probability that a given gene in a given individual will mutate is 0.005, and the effect of each mutation is to reduce survival by 1%. Mortality is applied in the diploid phase. In terms of the theory

will represent a small fraction of the whole, and if the population is relatively small it must also comprise only a small absolute number of individuals. This number will not be constant, however; merely by sampling error it will fluctuate from generation to generation. Sooner or later, it will reach zero. The optimal category is now that with one more mutation. If the population is sufficiently small the situation cannot be restored by back-mutation, since specific back-mutations will be extremely rare, and would have to occur in one of those individuals in the presently optimal category. Since this category in turn comprises only a few individuals, it will in turn become extinct long before the appropriate back-mutation occurs. As Muller put it, finite asexual populations incorporate a sort of ratchet mechanism, and can never bear fewer deleterious mutations than are currently present in their least-loaded line. They are therefore doomed to progressive loss of adaptation and eventual extinction.

Finite sexual populations might avoid this deadly ratchet. Gametes produced by recombination will vary, some bearing many deleterious mutations and others very few. The zygotes produced by fusion between heavily-loaded gametes will have low fitness, and their death will rid the population of many deleterious mutations simultaneously. Although the optimal class may disappear from time to time through sampling error, it can always be restored by fusion between lightly-loaded gametes. Sex can therefore restore the original vigour of the population, and make indefinite persistence possible. The mathematics of this process were

developed in section 10.3, the mutation rate per genome $U = 0.05$ and the selection coefficient $s = 0.01$, so that $U/s = 5$. Asexual individuals reproduce their own genotype, with reproduction following mutation and mortality. Among sexual individuals, the succession of events is mutation, mortality, meiosis with recombination, and gamete fusion. Recombination is simulated by placing a single crossover at random on the chromosome and recovering all possible gametic types. The population in the next generation is reconstituted by sampling 1000 individuals at random from an infinite population with the same frequency distribution as the previous population after selection. The upper diagram shows a sexual population, in which the optimal class (shown in black) fluctuates irregularly in frequency, sometimes disappears completely, but is always eventually reconstituted by recombination. The initial downward drift of the classes with 0, 1 and 2 mutations is caused by the small number of loci involved in the simulation. The lower diagram shows an asexual population, in which successive optimal classes bearing 0, 1, 2 ... mutations are irreversibly lost.

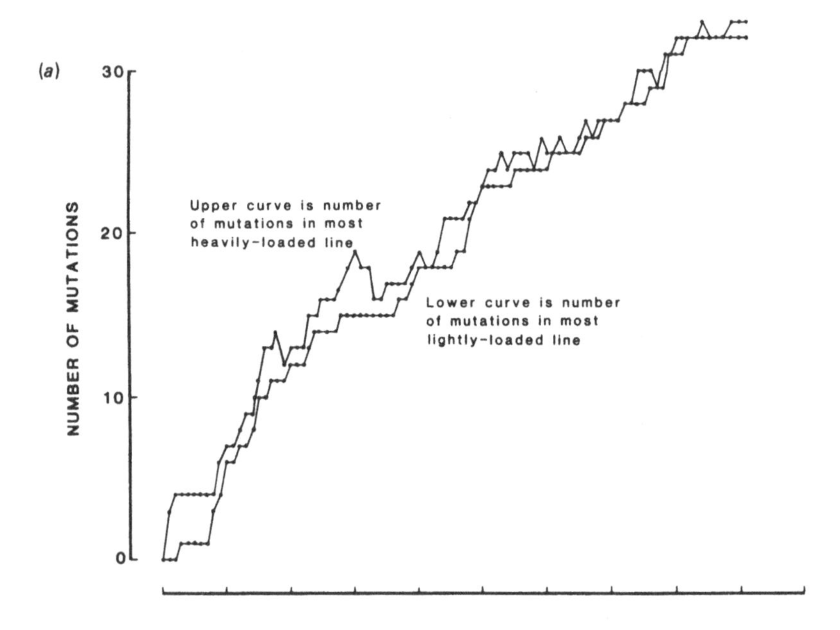
(a)
30
20
10
0
NUMBER OF MUTATIONS
Upper curve is number of mutations in most heavily-loaded line
Lower curve is number of mutations in most lightly-loaded line

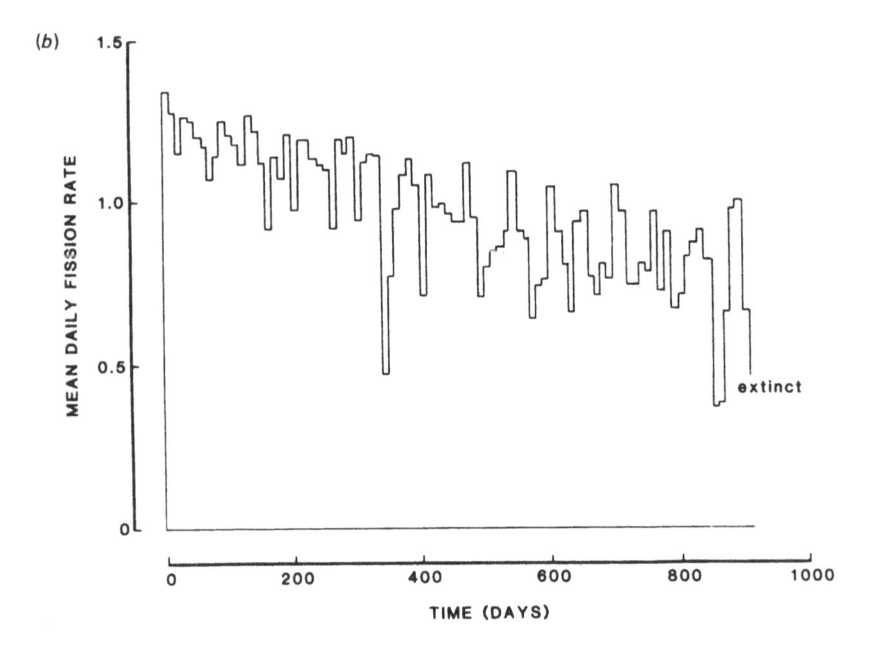
(b)
1.5
1.0
0.5
0
MEAN DAILY FISSION RATE
extinct
0
200
400
600
800
1000
TIME (DAYS)

worked out by Haigh (1978; see also Maynard Smith 1978), and it is described in more detail below in sections 10.3 to 10.5). The main feature of the process, the successive loss of the optimal class in finite asexual but not in finite sexual populations, is illustrated in Figure 40.

8.3 The Ratchet in isolate culture

The Ratchet operates only in finite populations, and, before going into details, it is intuitively obvious that it will move more rapidly in small populations – such as isolate cultures. Suppose that we set up a perfect isolate culture comprising a single line of an asexual organism that produces a single offspring and then dies. Since there is no selection, the rate at which this line decays will be a function of the mutation rate alone. I shall describe some experiments below which suggest that a reasonable estimate of this mutation rate is that one mutation which reduces viability by 1% arises every four or five generations. Within 500 generations, therefore, the vitality of the line will have sunk to about one-third of its original value, and extinction will be likely.

When more than one line is maintained, or more than one offspring produced, the problem is made more difficult by the action of selection in opposing mutation. In Figure 41 I have shown the output from a programme which simulates all the major features of a haplont

Figure 41. The Ratchet in isolate culture. These results were obtained from the programme simulating the fate of isolate cultures described in section 5.1. The system being simulated was a haplont whose genome comprises 100 mutable sites. Initial fission- and death-rates were 0.1 and 0.01 per hour respectively, giving an initial mean fission rate of about 1.3 per day; no fission was allowed for at least 10 h following a previous fission. There were four lines in the series. The mutation rate was $u = 0.001$ per locus and thus $U = 0.1$ per genome per generation; the selection coefficient was $s = 0.025$, so that $U/s = 4$. (*a*) The accumulation of mutational load. Note that the number of mutations in the most heavily-loaded line can either increase or decrease through time, while the number in the most lightly-loaded line can never decrease. The difference between the most heavily-loaded and the most lightly-loaded lines becomes smaller as the mean load increases, since the lines become less independent as the increasing death-rate necessitates more frequent transfers. (*b*) The associated trend in mean fission rate. The 10-day mean fission rate is plotted as a histogram to emphasize the similarity between this figure and the many published descriptions of senescence in isolate culture. Note the substantial stochastic variance around a downward trend. This series became extinct after 910 days and thus roughly 10^3 generations.

accumulating mutations while being maintained under the normal protocols of isolate culture. Although I have deliberately chosen a very low mutation rate – about one-quarter of the best estimate we have from measurements in *Drosophila* – mutations accumulate rapidly, the fission rate falls, and the series becomes extinct within 1000 generations.

There is no general difficulty, then, in accounting for the senescence of isolate series in terms of the Ratchet. However, the theory cannot be applied straightforwardly to ciliates, because of their peculiar cytogenetics. Since they have two kinds of nuclei they will be subject to two Ratchets, one accumulating mutations in the macronucleus over a sequence of asexual generations, and one operating in the germ-line micronucleus itself.

Granted that the macronucleus largely directs vegetative function, the clonal senescence of ciliates reflects macronuclear degeneration. The function of conjugation is then obvious, since the old macronucleus is destroyed and a new one constructed. Conjugation results in the demolition of the old structure and the erection of another from the micronuclear blueprint; Engelmann was right after all, in a sense.

Calkins (1920) was responsible for an ingenious attempt to test this hypothesis directly. During conjugation, mated paris of *Uroleptus mobilis* adhere only at their anterior tips, and the migratory pronuclei must pass through this bottleneck before fusion. After meiosis but before fusion, Calkins amputated the united anterior region, with the result that the macronucleus was broken down and reconstituted as usual, while the exchange of pronuclei was prevented. Unfortunately, very few individuals survived such radical surgery, but in those few both fission rate and clonal longevity were restored, just as they are by normal conjugation. The experiment is not conclusive, because the nature of the new micronucleus is unclear. If it were formed by the fusion of the two pronuclei, then it would be homozygous at all loci, since the two are mitotic copies of the same haploid meiotic product; the only surviving individuals would then be those which had few or no severely deleterious genes with low penetrance. However, it is as neat a demonstration as one could wish for of the macronuclear origin of clonal senescence. Further support is given by the observation (originally due to Dawson, 1919) that senescence occurs even in amicronucleate lines.

How does the macronucleus degenerate? I have already discussed the possibility that somatic assortment may lead to the loss of some essential genetic element. A second possibility, which I want to stress in this section, is that somatic assortment will also lead to homozygosity for mildly deleterious mutations, even when copy number is more or less conserved.

Macronuclear mutations may originate in either of two ways. In the first place, a mutation which occurs in a diploid micronucleus will constitute half of all the copies of that locus in the macronuclei of exconjugant lines which inherit it. If it is only mildly deleterious, nearly half of all the clonal descendants of these exconjugants will become homozygous for the mutation through somatic assortment. This will be an important source of mutations, since micronuclear genes are not expressed during vegetative reproduction, and mutations which are deleterious in the macronucleus will be neutral when they arise in the micronucleus. The micronucleus will therefore act as a rather efficient mutation accumulator,

Figure 42. Effect of *X*-irradiation on clonal longevity in *Paramecium aurelia*. Reanalysed from Fukushima (1974), Table 1. The analysis of variance for the effect of treatment on log clonal longevity is as follows.

Source	df	SS	MS	F	P
Between treatments	3	0.035805	0.011935	12.80	< 0.005
– linear regression	1	0.028602	0.028602	7.94	> 0.10
– deviations	2	0.007203	0.003602	3.86	> 0.05
Within treatments	8	0.007460	0.000933		
Total	11	0.043265			

The raw data are plotted in the figure. The linear regression shown is

$$\log_{10} t_E = 2.527 - 0.000909\ (R \times 10^{-3})$$

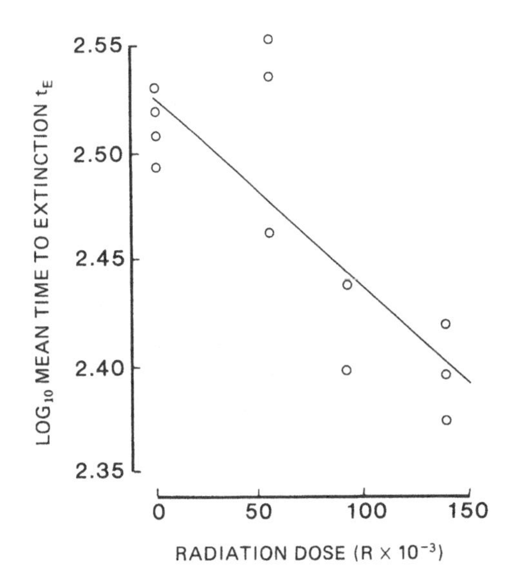

and this load will eventually be passed on to the macronucleus after conjugation.

Mutations will also occur in the macronucleus itself. If a mutation were to occur in any of the M copies of a locus, $1/M$ of all the lines eventually descending vegetatively from that mutant will be homozygous for the mutation. High copy number gives no protection against this process, since the number of mutations occurring will increase in direct proportion to the number of copies present.

Deleterious genes arising in either of these two ways will be made homozygous through assortment, and once homozygous for a mutation the macronucleus can never recover the wild-type allele. The irreversibility of this process means that any clonal isolate series will experience a Ratchet: without sex, no series can come to contain a line which is homozygous for fewer mutations than its currently least-loaded line. Since the total number of individuals involved is small – only about 20, at the most, in well-conducted experiments – the macronuclear Ratchet will turn rather quickly. The consequence of this will be the progressive and irreversible loss of vitality, culminating in extinction, that I have described at length in previous sections.

This explanation overcomes many of the difficulties encountered by the straightforward somatic assortment hypothesis. It accounts for the gradual loss of vitality in most cultures, for the parallel deterioration of the micronuclear genome, and for the occurrence of clonal senescence in organisms other than ciliates. It shares with its rival a scarcity of direct evidence. The obvious test would be to elevate the mutation rate, which should cause the Ratchet to move more quickly and would thus reduce clonal longevity. The best evidence I can find on this point was published by Fukushima (1974), who exposed exautogamous lines of *Paramecium aurelia* to three levels of *X*-irradiation, and measured their subsequent longevity and fission rate. There was no effect on fission rate or mortality for about 100 generations following the treatment, so any immediate physiological effect of radiation at the dosages used can probably be ruled out. Subsequently, fission rate declined more steeply in the irradiated groups than among unirradiated controls, as predicted by the Ratchet. Earlier work by Kimball and Gauthier (1954) looked for such an effect but failed to find it. Fukushima also gives the maximum lifespan of the control and irradiated lines, and I have reanalysed his data in Figure 42. Assuming single-hit damage, the elevation of the mutation rate should be directly proportional to the dosage of radiation administered. I shall show below (section 10.3) that the logarithm of time to extinction should decline linearly with the rate of mutation per genome. Thus, we expect a linear effect of radiation dosage on the logarithm of clonal longevity. The analysis shows a highly significant effect of treatment, but the evidence for linearity is not convincing. In my judgement failure of the linear model can be attributed to the relatively small size of the experiment (5 groups, including the control, with 2–4 series in each group). The remainder of Fukushima's results rather strongly support a mutation-accumulation hypothesis, as he himself points out.

The accumulation of deleterious mutations will cause an increase in variability. The appearance of various monstrous morphologies in aging cultures is of course well-known. More convincing evidence for a gradual increase in mutational load would be the increase in variation of a quantitative character under polygenic control. The only satisfactory data I have found in the literature is Frankel's (1973) study of ciliary row number in *Euplotes*, which I have reanalysed in Figure 43. There is clear evidence for increasing variability, as we would expect if the loci responsible were gradually accumulating an increasing load of deleterious

Figure 43. Increase of variability of ciliary row number with clonal age in *Euplotes*. Data from Frankel (1973): solid circles are subclones of K7 – 2 (his Table 4), open circles subclones of K7 – 6 (his Table 5). Plotted points are averages over subclones, ± one standard error. When these observations began the K7 clone was about 500 fissions past conjugation.

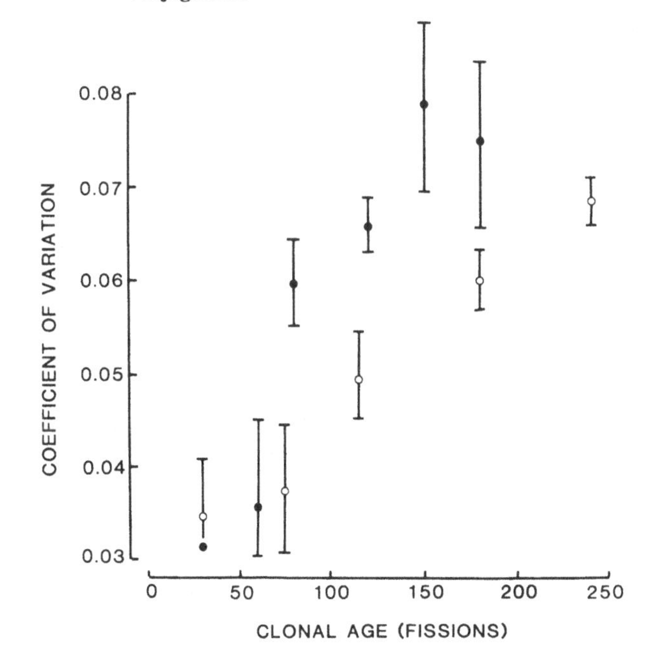

mutations. Among metazoans, Lynch (1985) found that substantial genetic variance for life history characters was created by mutation after 75 asexual generations in *Daphnia*.

Two points about this process should be made more explicit. The first is that it is not necessary – nor at all likely – that the *same* mutation or mutations be fixed in all lines. In the more general case, it is obviously unlikely that a deleterious mutation should be fixed in a reasonably large population. All that is required is that no line can acquire a smaller *total* number of mutations. Secondly, since amitosis will create variance in the number of mutations born by vegetative descendants, it might be imagined that selection could act on this variance so as to oppose the indefinite accumulation of mutations, as it does in the case of the variance created by outcrossing. The reason why this is not true is explained below, in section 10.6.

The history of isolate cultures of protozoans provides as comprehensive an illustration of the Ratchet as one could wish for. The amitotic macronucleus will become homozygous for deleterious mutations at a rate which depends on the rate of mutation and the rate at which the original mutant copy segregates. Without sex, no series can come to contain a line which is homozygous for fewer mutations than its currently least-loaded line. Because the number of individuals involved is small the macronuclear Ratchet will turn rather quickly. The consequences of this will be senescence, and eventually extinction. It is essential therefore, that an uncontaminated template be preserved, from which an undamaged macronucleus can be reconstructed. The micronuclear template, however, is itself being continually damaged by mutation. This effect will be a large one, in part because of the very small numbers involved, and in part because selection against deleterious mutations will be very weak since micronuclear genes are not expressed during vegetative reproduction. Sampling error will be a much stronger force than selection, and the micronuclear Ratchet will turn rather quickly too. The accumulation of deleterious mutations in the micronucleus is revealed by the startling effects of inbreeding aging clones. Their exposure to selection is shown by the segregation of large numbers of lethal, sublethal and subvital exconjugants. Sonneborn and Schneller (1955) refer to conclusive experimental evidence of massive mutational damage in old micronuclei, got by analysing crosses between young and old clones; but their report is only an abstract, and I have been unable to find a full account of this work. And finally, experiments such as those of Calkins (Figure 15) show directly how a population might be perpetuated indefinitely by occasional sexuality, despite the strictly limited lifespan of any of its asexual lines.

To sum up, conjugation rejuvenates ciliate lineages in two ways. In the first place, it is associated with the replacement of an old, Ratchet-damaged soma (macronucleus) from a germinal (micronuclear) blueprint. Secondly, it protects the integrity of the germ-line itself against a micronuclear Ratchet, bringing together deleterious mutations that have arisen in different lines of descent, so that the death of a few heavily-laden scapegoats can cleanse a whole community of its sins. However, the Ratchet has a significance far beyond the interpretation of long-forgotten experiments with ciliates, for it reflects problems that will arise whenever genetic information is transmitted. It is to these more general aspects that I shall now turn.

9

Soma and germ

9.1 The division of labour

A typical eukaryotic unicell has a dimension of about 100 μm and thus a volume of about $10^3\ \mu m^3$. Some are much smaller, and even approach prokaryote values of about $10\ \mu m^3$; others, such as large diatoms, amoebas and ciliates, are much larger and attain $10^6\ \mu m^3$ or more. The extent of this variation makes us wonder why some unicells are so much larger than others.

In modern systems, large unicells may escape the attentions of filter-feeders such as small cladocerans and rotifers. Grazers of about 0.1–1 mm in length usually have fairly well-defined upper limits on acceptable particle size, and cells greater than about 10^4–$10^5\ \mu m^3$ in volume will be virtually immune from grazing in most ecosystems. However, a more general advantage of large size may be the ability to resist periods of starvation. The absolute rates at which matter or energy are moved or used by organisms generally increase logarithmically with size, the exponent being less than unity, and commonly about 0.75 (Peters 1983); hence typically

$$\log_{10} R = \text{constant} + 0.75 \log_{10} m$$

where R is a metabolic rate and m is body mass. The specific metabolic rate therefore declines with size, since

$$\log_{10}(R/m) = \text{constant} - 0.25 \log_{10} m,$$

showing that the cost of maintaining unit mass of tissue is less for larger than for smaller organisms. The time required to exhaust a given quantity of reserves, expressed as a proportion of total body mass, should therefore increase with body mass. We anticipate that in starving poikolotherms

$$\log \text{survival time} = \text{constant} + \log_{10}(m/m^{0.75})$$
$$= \text{constant} + 0.25 \log_{10} m,$$

which has some experimental support from work on small crustaceans (Threlkeld 1976). Large size may thus be advantageous when the availability of nutrients fluctuates through time.

However, there is a corresponding, and much better characterized, disadvantage in getting bigger. The per capita rate of increase is a specific metabolic rate, and should therefore decline with body mass with an exponent of about −0.25. In particular, the rate of increase in numbers under nearly optimal conditions shows this relationship with body size

Figure 44. Allometry of the maximal rate of increase in small organisms. The thick line is the Blueweiss regression, extending over about 20 orders of magnitude from prokaryotes to large vertebrates, with a slope of about −0.25. The envelopes include all data points for the taxa named in capital letters; the crosses within circles are bivariate means for the taxa named in lower case letters. Note that colonial algae fall on the main regression, while algal unicells fall below it. The thick line within the envelope for algal unicells is the regression for these forms. The raw data on which this figure is based, with about 200 data points, are given by Bell (1986), who also presents an analysis of covariance showing that both multicellularity and nuclear dualism contribute to elevating the rate of increase at given size.

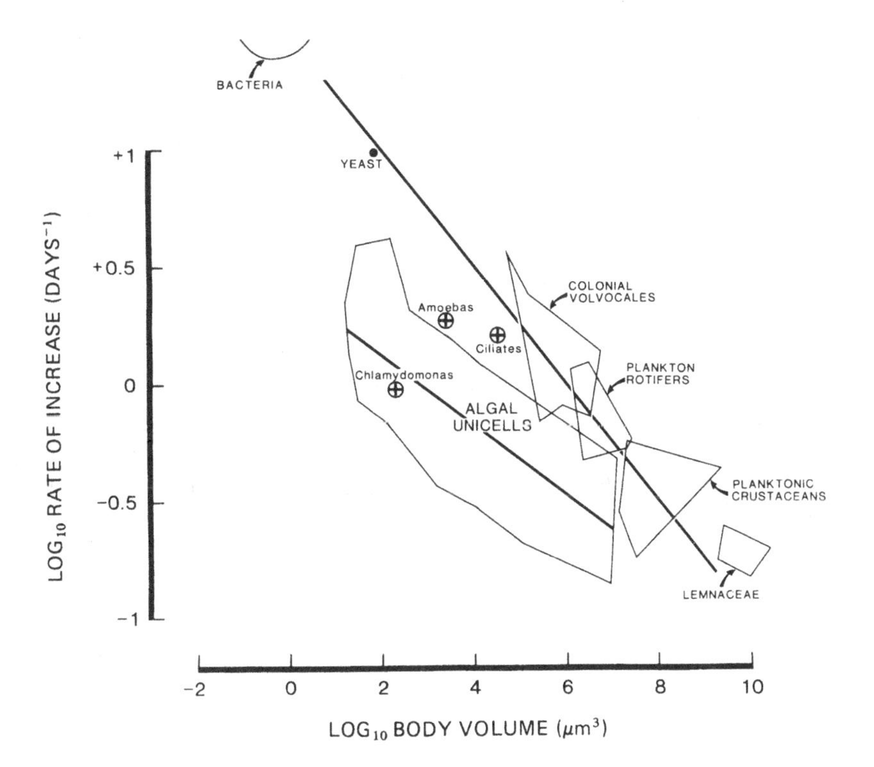

over 20 orders of magnitude from bacteria to whales (Blueweiss *et al.* 1978). For small organisms, with volumes of between 10^1 and 10^7 μm^3, the same general relationship holds (Bell 1985c; Figure 44), but some groups depart from the general trend. Ciliates and simple multicellular algae lie close to the general regression line, but unicellular algae, and perhaps amoebas, lie substantially below it. What is it about the construction of some organisms which enables them to increase in numbers much faster than others?

One mechanism underlying this trend may be that larger cells require more DNA and thus spend longer in cell division. DNA serves two functions: it supports genetic transmission by providing templates for further DNA synthesis, and it supports metabolism by providing templates for RNA (and thus protein) synthesis. Larger cells might be more complex and thus require more DNA for genetic purposes. On the other hand, larger cells might require more DNA to provide the RNA templates necessary to support their higher metabolism. In either case, we expect the quantity of DNA to increase with cell size. This prediction is eminently successful: when we analyse material as diverse as bacteria, unicellular algae, plant root-tip cells and vertebrate erythrocytes, we find that about 95% of the variance in DNA content is attributable to variance in cell size (Figure 45). The general relationship is even satisfactory for viruses, and thus extends over about 11 orders of magnitude. Over the whole of this range the exponent of the regression is less than unity, showing that specific DNA content (pg DNA/μm^3 cell volume) decreases with cell size. The value of the exponent varies taxonomically, being less for prokaryotes than for eukaryotes.

These data also enable us to discriminate between two proposed explanations of the increased DNA content of larger cells. Ciliate macronuclei, actively engaged in transcription but not involved in the sexual transmission of genes, fall along the main regression. Ciliate micronuclei, concerned with the specification and transmission of phenotype but not active in vegetative life, fall uniformly below the main regression, forming a separate group of points. The increase in DNA content with cell size amongst other organisms can therefore be attributed to the greater metabolic requirements of larger cells rather than to the greater genetic requirements of more complex cells. Considering ciliates alone, regression on cell size accounts for 97% of the variance in macronuclear DNA, while the small quantity (12%) of the variance in micronuclear DNA accounted for is not significant.

As an aside, the micronuclear genome may include, not only functional DNA specifying enzymes, but also 'selfish' DNA (Doolittle and Sapienza

Figure 45. DNA content and cell volume. Different taxa are indicated by: R virus, B bacterium, A unicellular alga, Y yeast, P angiosperm (root tips), V vertebrate (mainly nucleated erythrocytes), C ciliate macronuclei. Ciliate micronuclei are shown by the asterisks; values refer to the same species as the macronuclei. It has not been possible to display all of the 54 taxa. The regression equation, excluding micronuclei, is:

$$\log_{10}\text{DNA/cell} = -2.11 + 0.857 \text{ (se 0.028) log cell volume with } r^2\, 0.948,\ P < 0.0001.$$

The regression for micronuclei only is:

$$\log_{10}\text{DNA/cell} = -1.79 + 0.386 \text{ (se 0.417) log cell volume with } r^2\, 0.125,\ P = 0.39.$$

Analysis of covariance shows that the elevations of these two regressions are significantly different (at $P < 0.001$) over the range of values for ciliates. Source of data for DNA content is H. S. Shapiro's tables in the Handbook of Biochemistry; cell volumes previously collated by Bell (1985c).

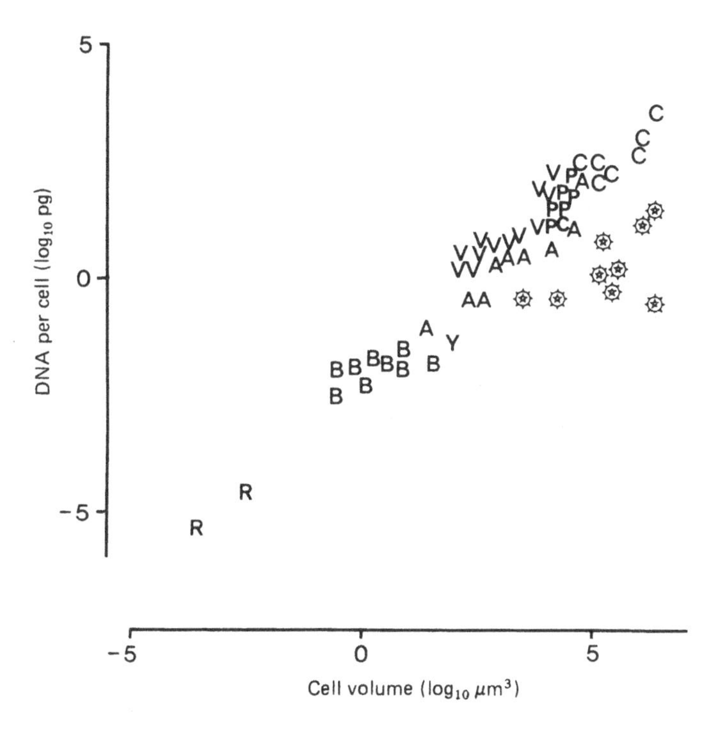

1980, Orgel and Crick 1980) which persists and spreads because it directs its own synthesis, but does not contribute to the phenotype of the organism, except as a consequence of sequestering the materials and time required for DNA synthesis. Since almost all the variance of macronuclear DNA is associated with differences in cell size, only a very small residual quantity (measured by the mean square of deviations from the regression line, $MS_D = 0.018$) is a candidate for selfishness. This is as expected: selfish DNA would have no interest in being transmitted to macronuclei – indeed, this would lower the fitness of selfish sequences, by reducing the rate of division of the parent cells. Among micronuclei, however, the residual variation is much greater ($MS_D = 0.563$). It seems reasonable to interpret the difference, $0.563 - 0.018 = 0.545$, as arising from the variation between species in the content of selfish DNA in the micronucleus. This is not much less than the overall variance of micronuclear DNA among species (0.551). Since DNA content varies between about 0.14 and 25 pg per nucleus, it appears that in many species the great bulk of the DNA is selfish in nature. This speculation is given some point by the observation that among hypotrichs the greater quantity of DNA eliminated during the formation of the macronucleus appears to consist largely of repetitive sequences.

The greater quantity of metabolic DNA required by larger cells will reduce the rate of increase, since the duration of both mitosis and meiosis increases with nuclear DNA content (van't Hoff and Sparrow 1963, Evans *et al.* 1972, Bennett 1971). Thus, the division time of cells increases, and their division rate falls, with size in part because of the requirement of larger cells for more DNA. The evolution of larger size is thus contingent on the evolution of mechanisms to handle large amounts of nuclear DNA. Above a volume of about 100 μm^3, this is achieved by the eukaryotic solution of chromosomes linearized by telomeres and moved around by centromeres and microtubules. Larger eukaryotic cells generally contrive to maintain an adequate quantity of DNA by becoming multinucleate, polyploid or polytene. Cells, especially free-living cells, of more than about 10^5 μm^3 in volume are rare, and most – for example, the giant multinucleate amoebas – are very slow-growing. The exceptions are the ciliates, which, despite their large size, achieve rates of increase comparable with those of multicellular organisms. The invention which enables them to do this appears to be nuclear dualism. The massive quantity of DNA required for vegetative function is sequestered into a macronucleus, which evades the usual penalty of a longer mitotic cycle through dividing rapidly by amitosis. Meanwhile, the germinal material, in which accurate mitotic replication is essential, is kept small in bulk and can divide rapidly also.

The evolution of large size among ciliates, then, hinges on a division of labour between soma and germ line, in this case between a somatic macronucleus and a germinal micronucleus. Still larger organisms achieve the same result by the same means, except that labour is divided between somatic and germinal cells rather than nuclei. There is no doubt that this mode of construction is successful. Colonial algae such as *Volvox*, despite their large size, lie on the Blueweiss line and thus have much greater rates of increase at given size than their unicellular relatives. In some way, the evolution of multicellularity reduces the penalty which unicells pay for size increase.

I have suggested that the crucial advantage of multicellular organization arises from the division of labour between soma and germ (Bell 1985c, which should be consulted for the quantitative details of the arguments in this section). The rate of uptake of nutrients will depend on the concentration gradient at the cell surface. The organism can usually do nothing to alter the external concentration (except by moving away from regions which it has depleted), but it can steepen the gradient by reducing the internal concentration. This can be achieved by sequestering the nutrients which have been captured, so that their presence no longer interferes with further uptake. Unicells can manage this to a limited extent with storage organelles such as starch-bearing pyrenoids. In multicellular organisms a more drastic and effective solution is possible: one whole category of cells, the germ cells, becomes specialized as a nutrient sink, while the remaining, somatic, cells act as a source. The translocation of substances from somatic source to germinal sink maintains a steep concentration gradient at the surface of the organism and thus increases the average rate of uptake.

A second fundamental advantage of multicellular organization arises from communication between somatic cells. Any cell concerned with nutrient uptake will experience some degree of variation of the external concentration from time to time. It can be proven that, for any given mean concentration, the mean rate of uptake will fall as the variance of external concentration increases. The average rate of uptake can therefore be raised not only by systematically maintaining a steep concentration gradient, but also by reducing the variance of the difference between external and internal concentrations. An isolated unicell has no means of doing this, but a multicellular organism can reduce the variance of internal concentrations by sharing nutrients between somatic cells. In this way the somatic cells of a multicellular organism will experience a less variable environment than an equivalent number of isolated unicells, and will acquire a greater quantity of nutrients in unit time.

The Volvocales, a family of green algae which includes both unicells such as *Chlamydomonas* and colonies such as *Volvox*, furnishes some particularly clear illustrations of the principles underlying the evolution of multicellularity. The lowest grade of colony organization is represented by loose and variable aggregates of cells which can gain no physiological advantage from being grouped together, and indeed are probably inferior to unicells because of their lower surface/volume ratio. Their greater bulk, however, will give them some measure of protection against grazing by filter-feeders. The colonies of genera such as *Stephanosphaera* and *Platydorina* are no larger, but consist of a fixed number of cells regularly arranged. These represent the reduction in the variance of form that one would expect to be the outcome of the stabilizing selection arising from two opposed selection pressures, the physiological disadvantages of undifferentiated aggregates of cells and the reduced grazing pressure and greater resistance to starvation associated with increased bulk. At this level of organization there is no evidence that multicellularity confers any physiological benefit: small algal coenobia consisting of ten or so cells arranged edge-to-edge have the same rate of increase, or perhaps a rather smaller rate of increase, than unicells of comparable size. The first real advance is shown by forms such as *Eudorina*, in which 16–32 cells surround a central cavity, the colony lumen. This arrangement has the straightforward effect of increasing the bulk presented by a rather modest protoplast volume. More profoundly, it represents the first physiological advance on the unicellular condition; substances can be moved from the somatic cells into the lumen, steepening the concentration gradient at the surface and enabling the lumen to be used as a homeostatic organ. *Eudorina* seems capable of substantially greater rates of increase than unicells with the same protoplast volume. There is still no division of labour between cells: all the cells in the colony are first somatic and then differentiate simultaneously into sexual or asexual germ cells. The distinction between soma and germ first arises in certain species of *Eudorina* and, more markedly, in *Pleodorina*. These are relatively large colonies, with 64–256 or more cells, and are characterized by a distinction between anterior somatic cells and a variable number of specialized germ cells posteriorly.

The appearance of a distinct division of labour beyond some critical colony size reflects an economic principle dating back to Adam Smith: it becomes profitable to divide labour when the market is sufficiently large, and the degree of division of labour will be proportional to the extent of the market. This concept is familiar, in a rather vague way, to most biologists: the number of distinct cell types, for instance, increases with the total number of cells in the bodies of metaphytes and metazoans, and only the larger colonies of termites and ants support morphologically distinct castes. In the large colonies of *Volvox*, which generally comprise about 10^4 cells, differentiation has proceeded further, so that the great majority of cells are now small somatic cells incapable of reproduction, while a minority are much larger germ cells. These germ cells, moreover, are now located in the colony lumen, so that by translocating substances into the lumen fluid the somatic cells can maintain a steep concentration gradient while providing a homeostatic environment for the germ cells. The climax of Volvocacean development is reached by the section *Euvolvox*, in which the somatic cells are connected to each other and to the germ cells by thick cytoplasmic strands. The greater ease of communication between cells effectively increases the extent of the market – again, an idea due to Adam Smith – and further enhances the advantage of dividing labour; the passage of nutrients between somatic cells along these cytoplasmic connections reduces the variance of internal concentration that each experiences, and at the same time the connections will allow the somatic cells to drain more effectively into the germ cells.

The interdependence of somatic cells and the division of labour between soma and germ create definable physiological advantages. In particular, phosphorus, which often limits algal growth in fresh water, is taken up from solution more effectively by colonies than by unicells. The colonies produce a greater mass of smaller propagules than unicells of comparable size, and in consequence have greater rates of increase. In this way, the penalty that unicells must pay for size increase is evaded in multicellular organisms by the creation of a separate caste of germ cells and the consequent sterilization of the soma.

My student V. Koufopanou and I have carried out an experiment to test the validity of these speculations (Koufopanou and Bell unpublished). In *Volvox carteri* the cells are connected to each other only by an extracellular glycoprotein matrix, and can be isolated without causing any apparent damage. When young spheroids bearing uncleaved gonidia are disrupted in a tissue homogenizer, some gonidia become completely separated from somatic cells, while others are still associated with more or less extensive sheets of somatic tissue. These two categories of germ cells, completely isolated and semi-isolated, can then be compared with germ cells borne by intact spheroids. We observed the growth of all three categories in a series of nutrient media varying in dilution. In highly concentrated media, feedback inhibition is likely to be strong, and there will be a large premium on dividing labour between source and sink, so that gonidia in intact spheroids will grow more rapidly than isolated gonidia. In highly dilute media this advantage will disappear since little if any feedback inhibition is expected. We can predict, therefore, that when the growth of the germ cells is measured there will be a significant effect of state (isolated gonidia contrasted with those borne by intact spheroids, the latter growing more rapidly), and, more importantly, that there will be an interaction between state and medium concentration, attributable to the greater difference between states at higher nutrient levels.

The results of the experiment are shown in Figure 46. The gonidia, whatever their state, grew faster in more concentrated medium; the presence of this environmental effect is, of course, necessary for the hypothesis to be tested at all. The semi-isolated gonidia, growing together with disrupted sheets of somatic tissue, grew at the same rate as the completely isolated gonidia, so that the mere presence of disaggregated somatic tissue does not contribute substantially to the growth of germ cells. Conversely, the gonidia within the intact spheroids grew faster than either category of isolated gonidia at all concentrations, demonstrating

Figure 46. The growth of germ cells inside and outside the soma in *Volvox*. See text for description of experiment. SVM is the degree of dilution of standard *Volvox* medium. (*a*) Absolute growth rates of gonidia at different levels of nutrient dilution: ○ within intact soma; ◆ with disorganized sheets of somatic cells; ● completely isolated from somatic cells. Analysis of covariance (see Koufopanou and Bell unpublished) shows that main effect of dilution and interaction between state and dilution are highly significant. (*b*) Specific growth rates (growth per unit mass) of gonidia. Note that isolated gonidia are more efficient at low nutrient concentration but less efficient at high concentration.

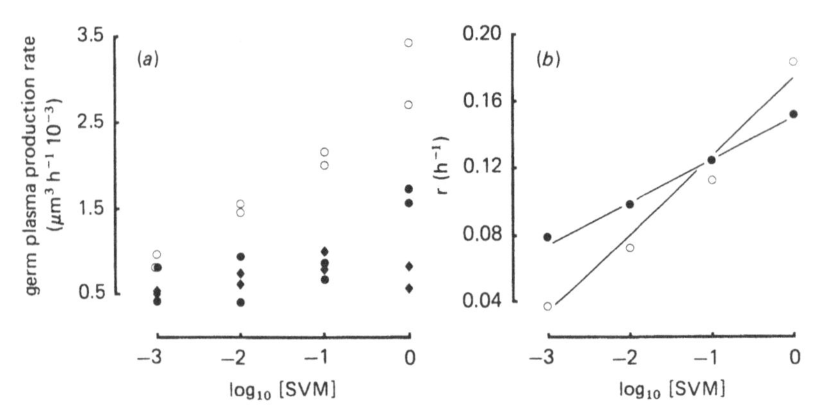

the advantage of a multicellular construction. The crucial observation is that the difference between the intact-spheroid gonidia and the isolated gonidia varies with concentration: there is little difference at low nutrient concentration, but at the highest concentration the gonidia within intact spheroids grew about three times as fast as the isolated gonidia. This is precisely the effect predicted by the source-sink hypothesis.

It might be argued that the process of isolation in some way damages the gonidia, so that the experiment merely verifies that undamaged cells grow faster. However, when we calculate the specific growth rate (growth rate per unit mass of tissue), we find that this is greater for isolated gonidia at low nutrient concentration but greater for intact-spheroid gonidia at high concentration (Figure 46 (*b*)). It is difficult to imagine how damaged cells could be more efficient than undamaged ones. Moreover, this transformation of the data suggests that either the unicellular or the multicellular modes of construction might be superior, depending on circumstances, with unicells having higher rates of increase at low external nutrient concentrations while the multicellular forms increase faster when nutrient concentration is greater. This is certainly consistent with the general observation that multicellular and colonial forms are most abundant in eutrophic systems (Reynolds 1984). It can be tested more rigorously using studies in which series of small ponds have been treated with different levels of mineral or organic fertilizers. The eutrophication caused by these treatments can be measured in a number of ways, the two most common being the concentrations of chlorophyll or reactive phosphorus. The response of any organism or group of organisms can then be expressed as the regression slope of their abundance plotted on the measure of eutrophication, with a steep slope indicating a large response to the treatment. The value of the slope in the most extensive data set we have found (DeNoyelles 1971) is shown in Figure 47 for different modes of construction within the Volvocales. The slope becomes progressively steeper as we pass from unicells to flat colonies with no differentiation between cells, to spheroids with various degrees of differentiation, and finally to spheroids with complete division of labour between soma and germ. This is not an isolated result; the less complete data given by Januszko (1971) for a similar experiment also shows the multicells responding much more steeply to eutrophication than do the unicells. Both our laboratory experiment and these field experiments, therefore, show how the physiological consequences of separating a somatic source from a germ-line sink are translated into the elevated rates of increase of simple multicellular organisms. There is then no bar to the evolution of much larger size and much greater elaboration.

9.2 Somatic and germinal senescence

The separation of soma from germ at once creates a life cycle in which the soma must be reconstructed in every generation from a germinal blueprint. It is a brute fact that the soma senesces, while the germ does not, in the short term. It is obvious why the germ line should not senesce – though not, of course, obvious how it should contrive to avoid doing so – but why should the soma grow old and die?

I think that the correct answer to this question was given by George Williams in 1957 (see also Williams 1966), following up a suggestion by Medawar (1952). Imagine two organisms, one of which produces, say, 10

Figure 47. The response of volvocacean algae to eutrophication in relation to their mode of organization. Plotted points are the slopes of increase in numbers of eutrophication for all members combined of one of the four different categories of organization. The two measures of eutrophication used are: ◆ chlorophyll *a*, and ▲ total reactive phosphorus. The classification of taxa into four modes of organization is due to Kochert (1973): 0. unicells; 1. flat plates of 4–16 cells with no lumen or division of labour; 2. forms with more than 16 cells, with a lumen and varying degrees of division of labour; 3. forms with several hundred or more cells and complete separation of soma from germ. Data abstracted from DeNoyelles (1971).

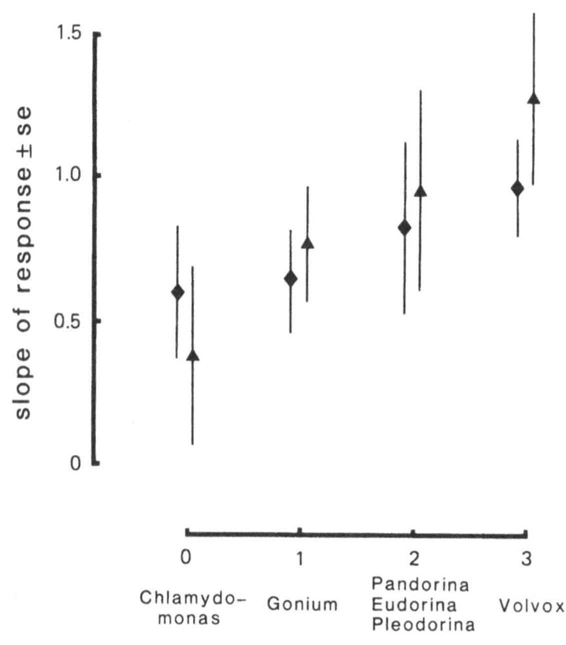

offspring at one year of age while the other produces the same number of offspring at two years of age. The former type clearly has an advantage, since when two years have passed its offspring will themselves have reproduced, giving rise to a lineage of 100 individuals, while the latter type has produced only ten. In the same way, alternative alleles will often be selected according to the age at which they are expressed, even though they have, in other respects, equivalent effects; and in general alleles which increase fecundity early in life will be selected more rapidly than alternative alleles which have the same effect on fecundity later in life. It follows that an allele which increases early fecundity may be favourably selected, even though it causes a reduction of fecundity or viability later in life. Indeed, the importance of early reproduction is such that an allele causing a slight increase in early fecundity may be favourably selected, eventually to become fixed in the population, even though it causes sterility or death later in life. It is, in fact, quite likely that alleles will exhibit pleiotropic effects of this sort, since an increase in present fecundity will often reduce subsequent fecundity by lowering the level of stored reserves, or decrease viability by exposing individuals to greater risks of predation. The empirical evidence for such 'costs of reproduction', and thus for the superiority of Williams' theory over its rivals, is reviewed at length elsewhere (Bell and Koufopanou 1986).

This theory emphatically does not imply that senescence is favoured as such. There is no suggestion, for instance, that senescence serves to eliminate old and worn-out individuals, as Weismann long ago speculated. Senescence rather arises as a more or less undesirable by-product of selection for increased vigour early in life.

It is a theory which applies specifically to multicellular organisms in which there is a clear distinction between soma and germ. It does not apply to unicells, as Williams himself recognised. With no distinction between soma and germ, there can be no selection leading to senescence, and therefore no life-cycle. To put this another way, Williams' theory derives its force from an unequivocal distinction between parent and offspring, with increased production of offspring leading to a reduction of parental vigour later in life. With equal binary fission, parent and offspring cannot be distinguished: both have the same age, and there is no way in which a gene acting before fission can exert a systematic pleiotropic effect on one of the cells resulting from fission. When I compared the life histories of oligochaetes reproducing by a more or less equal binary fission with those of other asexual invertebrates reproducing by eggs, I found that only the egg-producers showed a senescent decline in survival rate (Bell 1985a).

This argument implies that the parallel between somatic and germinal aging is unsound, since senescence will evolve as a byproduct of selection for increased early reproduction in the soma but not in the germ-line. Nevertheless, it is one which is often drawn, and I suspect that it lies behind the notion that somatic aging is caused by the innate decay of individual cells. It is this belief which has led to the enormous interest in cultured metazoan cells as a 'model' for whole-organism senescence.

Running parallel to the Ratchet is the theory that aging is due to the accumulation of somatic mutations, as advanced by Szilard (1959) and criticized by Maynard Smith (1959). That mutations of large effect (such as those which eventually give rise to a tumour) may kill organisms is not in dispute; rather, the general occurrence of senescent decline is interpreted as the gradual accumulation of many mutations of individually small effect, with death resulting when there are too few cells competent to carry out some vital function. The senescence of the metazoan body is then analogous with the senescence of a protozoan clone, except that the cells of the body, unlike those of the clone, do not have independent fates, the failure of one group of cells being capable of causing the death of the whole. This lack of independence, coupled with our ignorance of somatic mutation rates, makes it difficult to evaluate the plausibility of the theory. For example, if the somatic cells of a mammal behave like the *Drosophila melanogaster* germ line, they will accumulate about one mutation of small effect in every two cell generations. Even large mammals will go through no more than 40 cell doublings, accumulating an average of 20 mutations per cell. These will almost all be heterozygous, so that each will reduce cell viability by about 0.01. Their combined effect, assuming multiplicative action, will thus be to reduce viability to about 80% of its original value. However, the number of mutations per cell will have a Poisson distribution, with a variance equal to the mean. If the mutations do not effect the ability of the cells to divide, about 1% of cells will bear 33 or more mutations, and their viability will be only 70% or less of the original value. Such calculations are worthless, however, if a very small number of cells with a very large number of mutations have a disproportionate effect on the state of the organism, or if calendar time rather than cell-generation time governs the rate of somatic mutation, or if the somatic cells of mammals do not behave like the germ cells of fruitflies. The problem is best approached experimentally, which leads to issues too far from my main thesis to be discussed here. My own opinion is strongly influenced by the observation that most organ systems appear to senesce at about the same rate, with death resulting from any of a wide range of possible failures. This suggests that senescence is not due primarily to the accumulation of mutational damage, but rather to the generally decreased efficiency of damage repair systems later in life. This is readily explained by Williams' theory of the life history, in which genetic changes which decrease the efficiency of repair late in life will be favourably selected if they have the pleiotropic effect of increasing vigour early in life. This theory immediately predicts the more or less simultaneous senescence of different organs, since the net selection on any pleiotropic genetic change affecting a particular organ system will depend on the difference between its earlier and later effects on function. If a particular system ages more rapidly than all others, its failure will constitute a major source of mortality, and selection will favour mutations which postpone its senescence, even if they reduce function early in life. Once all systems fail at about the same age, all will evolve earlier ages of failure as the price for increased function early in life, until net selection is zero. This process is quite unlike the steady maladaptive deterioration of a clone under the Ratchet.

In short, the analogy between somatic and germinal senescence seems to be just an analogy. The germ line ages because finite asexual populations will inevitably accumulate mildly deleterious genes; the soma ages because selection will actively favour genes which cause loss of function in old organisms, provided that they are associated with early reproductive success. It is unlikely, therefore, that understanding the immortality of the germ line will tell us much about the mortality of the soma.

9.3 The life-cycle of ciliates

A clone will not age as a metazoan soma does, because protists of the present generation cannot enhance the reproduction of their remote ancestors, now nonexistent, by sacrificing themselves. The existence of a life-cycle in asexual protists therefore represents a serious challenge to current evolutionary theory.

There is no reasonable doubt that such a life-cycle exists, at least in ciliates. Having first been described by Maupas, a succession of qualitatively different stages has been recognized by almost all subsequent authors. These stages are typically an initial period of immaturity, during which conjugation does not occur; a period of maturity, during which mating occurs readily, and leads to the production of viable exconjugants; and a period of old age, signalled by the decreasing vitality of exconjugants and eventually, vegetative senescence. I have interpreted the later stages of this process as being caused by the irreversible accumulation of deleterious mutations in isolate lines. They do not, therefore, represent an adaptive process of any kind, but merely reflect intrinsic limits to the fidelity with which genetic information can be transmitted. This opinion runs directly counter to that of many previous authors (see Siegel 1970), who interpret senescence as an integral part of an adaptive sequence of change. But how, then, are we to interpret the earlier stages of the life history, and in particular the existence of a period of immaturity? This cannot reflect a trade-off between earlier and later reproductive success, since the individuals which represent the earlier generations cannot be influenced by the behaviour of their descendants; nor is it reasonable to interpret the acquisition of sexual competence as an outcome of genetic decay.

I think that the crucial observation, well-known to ciliate biologists (e.g. Nanney 1980), concerns the variability of the different stages. The onset of senescent change is highly variable, suggesting the operation of the sort of stochastic processes I have discussed above. The onset of sexual maturity, on the other hand, is rather invariant, with independent cultures of the same strain becoming mature at almost exactly the same time. This time differs between strains, and is readily altered by artificial selection (Sonneborn, 1938). This suggests to me that the occurrence of a period of immaturity is an adaptive response to the nonadaptive process of senescence.

Both micronucleus and macronucleus degenerate in strictly asexual lineages, and so both sexual and vegetative performance deteriorate through time. Vegetative senescence implies that the clone will eventually become extinct, so that nuclear reorganization at some stage will certainly be favoured by selection. Sexual decay implies that this reorganization cannot be too long delayed; there will be a severe penalty for long periods of immaturity which push back sexual expression so far that the sexual products are largely inviable. This is why the period of immaturity is not too long. On the other hand, any period of reorganization, by conjugation or automixis, is expensive in terms of time, materials and energy; mating, meiosis and macronuclear dissolution and reconstruction combine to slow down reproduction. One can summarize the effects of reorganization, therefore, in terms of forfeiting a number of fissions which could otherwise have been achieved. The shorter the period of immaturity, the more frequent conjugation will be, and the more such forfeits must be paid in unit time. This is why the period of immaturity is not very short.

The observed period of immaturity is then a compromise between a longer period that would reduce the viability of exconjugants too far, and a shorter period that would entail too great a loss of opportunities for reproduction. This interpretation is somewhat similar to the 'mutation-accumulation' theory of life histories proposed by Edney and Gill (1968), following initial suggestions by Medawar (1952). This is still not conclusively falsified as a theory of aging in metazoans, though current experimental results have tended to reject it in favour of Williams' theory of negative pleiotropy (Rose 1983; Bell and Koufopanou 1986).

To see how an intermediate age at maturity will maximize fitness, imagine a ciliate in which the rate of fission declines linearly through time with slope $-m$. Conjugation immediately restores the original rate of fission B_0, but prevents any fission from occurring for a period of C days. For simplicity, all exconjugants are assumed to survive, irrespective of clonal age. This scheme is illustrated in Figure 48. What is the fitness of a clone which conjugates at age A? This will depend on the number of exconjugant lines descending from a single founding cell, which is equal to $2^{B_0} \times 2^{B_1} \times 2^{B_2} \times \ldots \times 2^{B_A}$, where B_t is the number of fissions achieved on day t. The logarithm of this number is the sum $(B_0 + B_1 + B_2 + \ldots + B_A)$, which is equal to the area of the quadrilateral with base A in Figure 48. By simple geometry, this area is $A(B_0 - \frac{1}{2}Am)$. Since this occupies a period of time $(A+C)$, the average rate of fission is $A(B_0 - \frac{1}{2}Am)/(A+C)$. By differentiating this quantity, we find that the value of A which maximizes the mean fission rate is the value A^* satisfying

$$-\tfrac{1}{2}m(A^*)^2 - mCA^* + CB_0 = 0.$$

We can neglect terms in m^2, since m is a small number. After B_0/m days

Figure 48. A simple geometrical model of the ciliate life history. The clone has an initial fission rate of B_0 per unit time, which decreases linearly through time with slope $-m$. Conjugation at age A restores the original fission rate B_0 after inhibiting fission altogether for a period of C days.

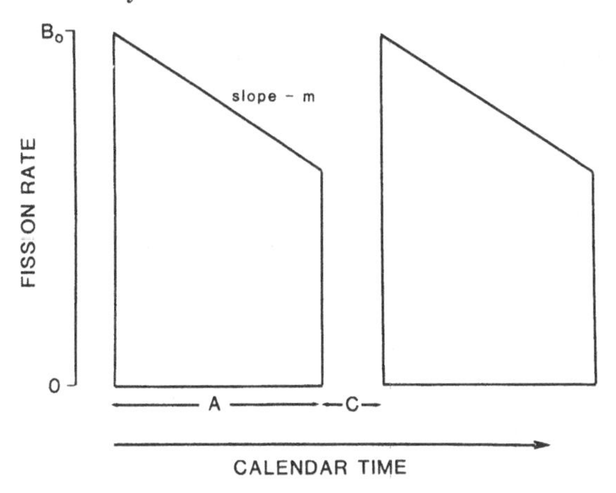

the fission rate will have fallen to zero; this will be roughly equal to the time to extinction of the line, t_E. We can then write

$$A^* = \sqrt{C}[\sqrt{2t_E} - \sqrt{C}].$$

as an approximation to the optimal value of A. The period of immaturity will thus increase with the square root of the lifespan of the line.

To illustrate this argument, I have used data on automixis in *Paramecium aurelia*. An additional complication is that the survival rate of the exautomicts declines with clonal age. Using the data of Pierson and Gelber shown in Figure 27, I find that the logarithm of the survival rate declines linearly with clonal age (in days), with a slope of $-k = -0.0086$, or roughly -0.01. The longevity of the nonautomictic lines described by Sonneborn (1954) was roughly 100 days, and 'fission rates usually drop markedly for one or more days at the time of endomixis' (Sonneborn 1937), so $C \approx 1$ day. We can incorporate the effect of exautomict mortality into the model by multiplying the number of vegetative descendants entering automixis by the survival rate of the exautomict lines. This yields

$$-\tfrac{1}{2}m(A^*)^2 - mCA^* + C(B_0 - k) = 0.$$

Ignoring terms in m^2 as before, we can rearrange this as:

$$(A^* + C)^2 = 2C(t_E - k/m).$$

Since in this case $t_E \gg k/m$, we have roughly

$$A^* \approx \sqrt{2t_E},$$

or $A^* \sim 14$ or 15 days as our prediction of the optimal length of the period of immaturity. The 200 or so measurements of the interendomictic period collated from seven papers by Sonneborn (1937) have a median value of about 20 days. For so crude a model, this seems quite a satisfactory agreement.

To sum up this section, I have attempted to interpret three phenomena: the life cycle of a metazoans, including senescence; the senescence of isolate lines of protists; and the life cycle of ciliates. Metazoan senescence is a byproduct of selection for an optimal life-history, which favours genes with beneficial effects early in life even though they cause grave damage to older animals. Protozoan senescence is the nonadaptive consequence of an irreversible accumulation of deleterious mutations under the Ratchet. The ciliate life history evolves as a consequence of senescence, with the period of immaturity representing a compromise between senescent decline favouring earlier maturity and the expense of nuclear reorganization favouring later maturity. These are three quite distinct processes, and few if any useful parallels can be drawn between them.

10

Mortality and immortality in the germ line

10.1 Endogenous and exogenous repair

All structures are damaged as time passes. If repair is imperfect, then deterioration is as inevitable as damage. Somatic repair is clearly imperfect, since individual senescence and death probably occur in all organisms where an individual can be unequivocally defined. But while individuals die, lines of descent persist, perhaps indefinitely; if this were not so, life itself could not have persisted through geological time. The repair of the germ line must therefore be much more effective than that of the soma. At the same time, the isolate culture of protozoans provides indisputable evidence that the germ line itself may irreversibly decay under certain conditions. To understand the relationship between somatic and germinal aging, and the distinction between mortal and immortal germ-lines, we must identify the mechanisms of germ-line repair.

Imagine that we are dealing with some human artefact, such as a chair. More precisely, we should speak of a system in which chairs are one component while the craftsman who makes the chairs is the other. The instructions given to the craftsman are that each set of chairs should correspond exactly, so far as that is possible, with an example of the set of chairs that he made previously. To prevent chairs from successive sets deteriorating into a haphazard pile of sticks and boards, two sorts of procedure might be used. In the first place, each chair might be scrutinized for any departure from the master chair, and any discrepancies made good. Secondly, a whole batch of chairs might be constructed, without repair, and each then subjected to some test of function. Each chair might be sat upon a thousand times by potential customers, and all those which eventually collapsed would then be rejected, the surviving chairs providing the master chair from whose design the next set of chairs is made. These two procedures represent fundamentally different systems of repair. The first is endogenous, in that the instructions for correcting errors reside within the system itself. The second is exogenous, in that these instructions reside, in part, outside the system. Living organisms possess endogenous repair systems of many different kinds, but all of them necessarily involve the comparison of a fabricated structure with some blueprint or master copy. The blueprint itself always consists of a sequence of bases in a nucleic acid molecule, although this will not normally be involved directly in the process of comparison. Endogenous repair is thus essentially a conservative process; it does not involve replication, and therefore operates principally in the soma. Exogenous repair systems work when variance in the number of errors per copy generated within the system is subjected to an external test; in living organisms, the test constitutes natural selection. The combination of variation and selection implies that exogenous repair systems operate only in replicating systems; they are therefore particularly important in the germ line.

The argument that I shall develop in the paragraphs below is that endogenous repair cannot be perfected, and must eventually give way to increasing deterioration and decay. The long-term survival of the germ line therefore requires exogenous repair systems, the most familiar of which is provided by sexual reproduction.

10.2 Recombination as a system for endogenous repair

The distinction between endogenous and exogenous repair is crucial because recombination can accomplish both. In an important series of articles, Bernstein *et al.* (1985a, 1985b, 1985c) propose that recombination acts primarily as an endogenous system of repair. Since both their theory and the Ratchet interpret sex as a repair device, it should be emphasized that they invoke totally different types of repair.

Bernstein *et al.* make a useful distinction between 'damage' and 'mutation'. 'Damage' is a non-replicated and therefore non-transmissible lesion of the DNA, caused, for example by breakage, cross-linking, or the insertion of modified bases. It can occur either in one strand or in both. It can be effectively repaired by endogenous systems because it can be identified unambiguously, and if repair fails it is usually lethal. Single-strand damage can be made good by excision-repair enzymes which use the undamaged strand as a template. Double-strand damage can be repaired only if an alternative template – an homologous chromosome – is available. Recombination between the damaged and undamaged homologues then provides a means of patching the damaged strands, using the undamaged strands as a template. Bernstein *et al.* make a thoroughly convincing case for the importance of recombination in repairing double-strand damage, which I accept. Whether or not this is the original function of recombination, as they claim, cannot be ascertained, but it is a very reasonable speculation. My chief reservation is that it would not be effective in asexual haplonts. Maynard Smith (personal communication) has pointed out that while the repair of double-strand damage requires recombination within the gene (or short section of DNA) it does not require recombination between flanking markers.

In dealing with the second category of error, mutation, the endogenous-repair hypothesis is much less convincing. A 'mutation' is defined as a change which leaves the physical regularity of the DNA molecule intact, and which is therefore stably inherited: base substitution is one example. The difficulty is, of course, that mutated bases or sequences cannot be unambiguously distinguished from their unmutated homologues, unless the mutation occurs when the original and daughter strands can still be told apart, for example by methylation. To resolve this difficulty, Bernstein *et al.* abandon the endogenous-repair hypothesis and propose instead that dealing with mutations requires outcrossing, so that the functional gene from one parent will mask a dysfunctional gene from the other. Closed sexual systems such as autogamy will not do this as effectively because many progeny will be homozygous for newly-arisen deleterious mutations. This is a perfectly reasonable argument, but amounts to no more than the conventional analysis of selection for outcrossing in terms of inbreeding depression. It has little relevance to endogenous repair.

It is important to note that an open genetic system is not required for endogenous repair: a closed system would work as well or better. Indeed, even meiosis would not be necessary in a diplont, if rates of mitotic recombination could be elevated. The endogenous-repair hypothesis therefore tells us little about the evolution of gametes or outcrossing, which are involved only in the exogenous repair made necessary by the Ratchet.

10.3 How fast does the Ratchet turn?

As with the craftsman and his chairs, each new generation of any organism is produced by using examples of the previous generation as blueprints. In asexual organisms the blueprint is adhered to very strictly, and if the copying process were perfect each generation would be identical with that preceding it, and there would be no accumulation of errors. It is, however, impossible, even in principle, to build a machine that will copy itself with perfect fidelity. Whenever information is transmitted, some part of that information will be irretrievably lost. In living organisms, the loss of information is represented by the transmission of uncorrected mutations. As I have described above (section 8.2), deleterious mutations accumulate through time in a finite asexual population, because the class with the least mutations must sooner or later be extinguished by sampling error, and cannot afterwards be restored. The main importance of the old observations of isolate cultures of protozoans is the evidence they provide for the operation of this Ratchet. But how potent is the process? Is the deterioration of the germ line likely to proceed so rapidly that asexual lineages cannot persist through geological time?

In an infinite asexual population, Haigh (1978) has shown that the frequency of individuals bearing a given number of mutations will follow a Poisson distribution. In a finite population this deterministic distribution will be only approximately realized, with the number of individuals in each class fluctuating stochastically from generation to generation. The rate at which adaptation is degraded will depend primarily on the time taken for the least-loaded class to fluctuate to extinction and this in turn will depend primarily on the absolute number of individuals in this class. If the total number of individuals in the population is N, then the number of individuals bearing the fewest mutations is expected to be

$$n_0 = N \exp(-U/s)$$

where s is the selection coefficient and U is the rate of mutation per genome per generation. Mutations are assumed to have multiplicatively independent effects on fitness. The Ratchet will then turn rapidly if n_0 is small. This will obviously be the case if N and s are small and U large. However, the meaning of these parameters needs to be clarified before we can draw any quantitative conclusions.

The Ratchet refers to a closed population, which is regulated at some given number N. It is essential that migration can be neglected, since it would otherwise provide a way in which the optimal class could be restored after its disappearance. It seems safest, therefore, to regard N as the total number of individuals in the species, although in a viscous species with many nearly isolated, independently-regulated local populations the appropriate value of N might be much smaller than this. In any real species, of course, N will vary, perhaps with large magnitude, from generation to generation, and it is not known how the appropriate mean value should be calculated. It seems likely, though, that it will be strongly influenced by the minimal population size, since, as we shall see, the probability of extinction per unit time decreases steeply with n_0.

The selection coefficient s takes some particular value, and the mutations to which the Ratchet refers are those which have this value, or those falling within a certain interval for which s is an appropriate midpoint. U then represents the number of mutations of effect s which arise per genome over one generation. Naturally, the narrower the range of selection coefficients to which s refers, the smaller will be the value of U. We might (and in practice we shall have to) take U to be the total rate of deleterious mutation, in which case s would be the value which correctly represents the accumulation of mutations over a very broad range of possible selection coefficients, but it is not known precisely how such an average should be calculated.

It follows from this interpretation that U and s will covary; mutations of small effect will be more frequent than those of large effect. Thus, a plot of U on s would be negative, with the intercept at $s = 0$ representing the rate of neutral mutation, and becoming asymptotic at some value of U for large s (ignoring lethals, which of course are not subject to the Ratchet). Integrating this curve will give the total rate of deleterious mutations per genome. I shall claim below that in *Drosophila* this total rate is about 0.4, including a few lethals; however, this estimate overlooks many mutations of very small effect, and the graph of U on s is not known for any case.

At one extreme are mutations of small effect ($s < 1/N$) but rather large frequency. For this class, U/s is large and therefore $\exp(-U/s)$ very small. Even in very large populations, the optimal class will usually consist of a single individual. The rate at which the mean load of the population increases will then depend on the rate at which optimal classes of one individual are successively eliminated, and replaced with the next least-loaded class, by sampling error. Unfortunately, this rate cannot be calculated analytically, and even a reasonable approximation is not easy to obtain. Haigh (1978) suggests treating the optimal class as if it were an isolated population, and then using the stochastic theory of branching processes to investigate its longevity. This involves assuming that the rest of the distribution behaves deterministically, so that the mean fitness of the population remains constant. This should be a reasonable approximation, provided that the optimal class comprises only a very small fraction of the population. The dynamics of this class then depend on the probability distributions of birth and death for its members. The probability of death is simply unity, since all its members will die at the end of each generation, to be replaced by their offspring. The probability of birth, in a population at its deterministic equilibrium, will be equal to the reciprocal of the population mean fitness. This is because the numbers of offspring in each class at the beginning of each generation must be divided by the total number of offspring produced by the population in order to maintain a constant population size. The mean fitness is $\exp(-U)$ for any selection coefficient s. At the same time, only a fraction $\exp(-U)$ will fail to undergo mutation, so that the mean birth-rate is $e^{U} \times e^{-U} = 1$. Hence the net rate of increase, as the difference between the birth-rate and the death-rate, will be zero. The only remaining parameter is the distribution of the number of offspring per female, which I have taken to be Poisson. The probability that a class initially comprising a single individual will become extinct within t generations is then

$$P(0, 1, t) = \exp[P(0, 1, t-1) - 1]$$

which must be solved numerically. The mean time to extinction of such a population depends on the upper limit N_{max} placed on the number of individuals in the population. With $N_{max} = 1$ the population cannot

Figure 49. The time taken for a cohort initially comprising a single individual to become extinct. The time to extinction, t_E, is plotted as a function of the maximum permitted population, N_{max}. Each point is the mean of ten independent replicates; three such series of ten replicates each were run. In each simulation, a population is allowed to propagate itself, assuming that the number of offspring per individual has a Poisson distribution with a mean of one, until it reaches a value of zero.

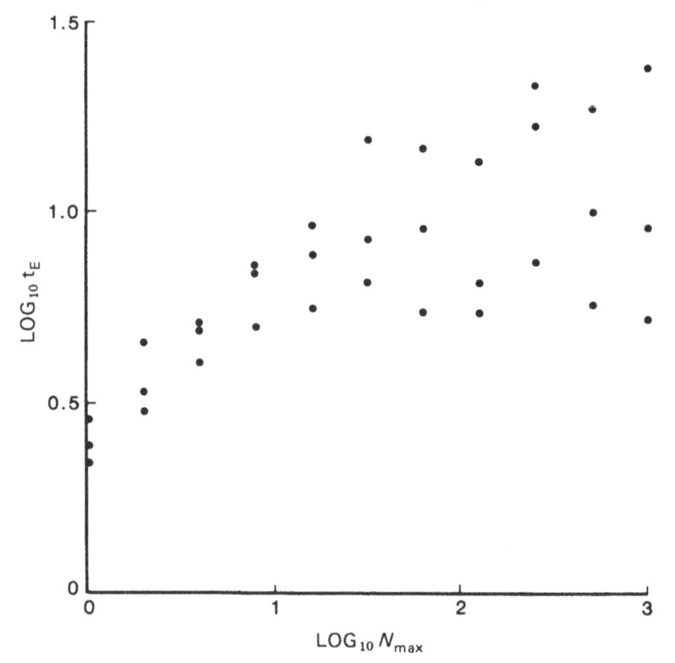

increase at all, and becomes extinct, on average, within two or three generations. This time increases to about 10 generations for $N = 50$, but then remains substantially the same for N_{max} up to 1000. This is largely because the distribution of times to extinction is skewed, with most populations becoming extinct very quickly, but a few lingering on for much longer periods of time. Figure 49 shows the relationship between the time to extinction and N_{max} found by numerical simulation of such simple stochastic populations. A useful bench-mark in thinking about the Ratchet is that, when selection and density-regulation are ignored, a cohort founded by a single individual is likely to disappear within about ten generations.

More generally, an optimal class of any size will become extinct within t generations with probability $[P(0, 1, t)]^{n_0}$ since all lineages are assumed to be independent. It will become extinct with probability Q within n_0 times the corresponding extinction time for a single individual, i.e.,

$$t(0, n_0, Q) = n_0 \times t(0, 1, Q),$$

I have again illustrated this point by a series of simulations (Figure 50). The graph of $\log t(0, n_0)$ on $\log n_0$ should have a slope of unity and an

Figure 50. The mean time to extinction of cohorts of different initial sizes. The maximum permitted size of the population is 1024 individuals; each point is the mean of 20 replicates.

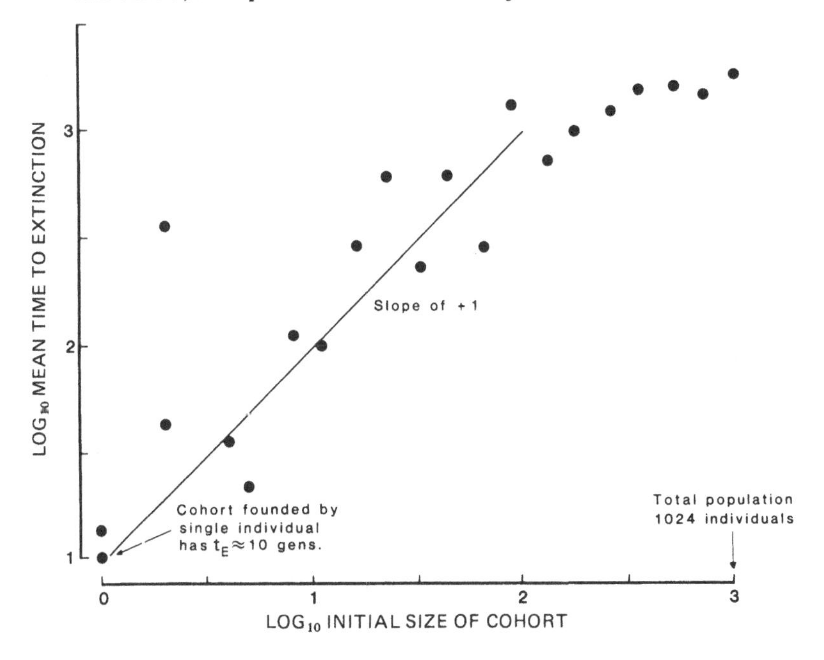

elevation (at $\log n_0 = 0$) of $\log t(0, 1) = 1$. The simulations show that this is the case, up to the point where the initial size of the cohort becomes a large fraction ($> 10\%$) of the maximum permitted size of the population.

These simple models fail to take into account the effect of the rest of the population on the rate of replacement of the optimal class. As the frequency of unloaded individuals increases, the mean fitness of the population increases, and the relative fitness of unloaded individuals must therefore decline, since only a fixed number of offspring are permitted to survive. To obtain a more accurate estimate of the rate at which the Ratchet turns, we must find out how the mean number of surviving offspring produced by unloaded individuals varies with their frequency.

Suppose that we take a population at its deterministic equilibrium, distinguishing between unloaded individuals on one hand, and all the loaded individuals on the other. We imagine that the frequency of loaded individuals is changed from its equilibrium value $\hat{P}_0$ to some arbitrary value P_0. The overall frequency of unloaded individuals must change in consequence, but the frequency of any loaded class remains the same relative to that of any other loaded class. The distribution is thus the same as the equilibrium distribution, except for the altered frequency of the unloaded individuals. If the frequency of the k-th class ($k \geqslant 1$) in the equilibrium population is $\hat{P}_k$, its frequency in the new population is $\hat{P}_k/(1-\hat{P}_0)$. The mean fitness of these individuals is

$$\sum_{k=1}^{\infty} \frac{\hat{P}_k}{(1-\hat{P}_0)}(1-s)^k = (\hat{w}-\hat{P}_0)/(1-\hat{P}_0),$$

where $\hat{w} = \exp(-U)$ is the mean fitness of the population at equilibrium. The total number of offspring produced by loaded individuals is therefore $N(1-P_0)(\hat{w}-\hat{P}_0)/(1-\hat{P}_0)$. Adding the contribution made by unloaded individuals, the total offspring production of the population is $NP_0+N(1-P_0)(\hat{w}-\hat{P}_0)/(1-\hat{P}_0)$. In every generation, this number is renormalized to N; therefore the mean number of surviving offspring produced by an unloaded individual, given that the frequency of unloaded individuals is P_0, will be

$$b_0 = \hat{w}(1-\hat{P}_0)/[\hat{w}(1-P_0)+(P_0-\hat{P}_0)].$$

This gives us the model we seek: b_0 is a decreasing function of P_0, since $\hat{w} < 1$.

We can therefore simulate the behaviour of the optimal class more realistically than before by allowing offspring production to vary in this fashion with population size, without setting any fixed upper bound on the number of individuals. This amounts to viewing the optimal class as an isolated density-regulated population. The change in the number of individuals per generation will be

$$\Delta n_0 = n_0(b_0 - 1),$$

which, provided that $\hat{w} \gg P_0$, is approximately equivalent to

$$\mathrm{d}n_0/\mathrm{d}t = cn_0(1 - n_0/\hat{n}_0),$$

where $c = (1-\hat{w})\hat{P}_0/\hat{w}$. This is a well-known logistic form. The parameter c is analogous to the limiting rate of increase r in the more familiar version

$$\mathrm{d}N/\mathrm{d}t = rN(1 - N/K),$$

used extensively in population dynamics. Therefore the value of c measures the strength of the restoring force which moves the population back towards $\hat{n}_0$ after it has been perturbed. If $\hat{w} \gg P_0$, as I have assumed, then c will be small. This implies that $\hat{n}_0$ will be approached asymptotically, without overshooting, from either above or below. It also implies that the restoring force will be weak, and convergence on $\hat{n}_0$ slow, unless n_0 is much larger or much smaller than $\hat{n}_0$, so that the density-independent arguments set out previously should provide a fairly good approximation unless the population is far from equilibrium.

The results of direct simulation of logistic populations regulated in this manner are shown in Figure 51. The relationship between the mean time to extinction t_{E} and the equilibrium population size $\hat{n}_0$ is found empirically to be:

$$\log_{10} t_{\mathrm{E}} = 0.982 + 1.055 \log_{10} \hat{n}_0.$$

This substantiates the conclusion from simpler models, that the log–log plot of t_{E} on $\hat{n}_0$ has a slope and a zero intercept of about unity. We can now bring all these numerical results together by saying that the number of generations elapsing before the optimal class vanishes is rather closely approximated by

$$\log_{10} t_{\mathrm{E}} = 1 + \log_{10} \hat{n}_0,$$

or more simply, $t_{\mathrm{E}} = 10\hat{n}_0$.

Once the optimal class has become extinct, the frequency distribution of load shifts backwards until a new equilibrium is approached. The rate of this process depends not on population size, but on the relative strengths of selection and mutation. At the time when the optimal class disappears, the next class, with one more mutation, has size n_1. According to Haigh (1978), n_1 is reduced to about $1.6\hat{n}_0$ in roughly

$$(\log s - \log U)/\log(1-s)$$

generations. We can thus distinguish two phases in the replacement of the optimal class by the next least-loaded class. The first is an 'establishment' phase, during which the newly optimal class drops from its initial nearly deterministic size of n_1 to about $1.6\hat{n}_0$ individuals; this process involves the backward displacement of the whole distribution, and is nearly deterministic. The second is an 'extinction' phase, during which a class of $1.6n_0$ individuals declines to zero; this is determined stochastically by sampling

Figure 51. The mean time to extinction as a function of the equilibrium population size in isolated logistic populations. The birth-rate declines with population size in the manner described in the text. Each plotted point is the mean of ten replicates. The regression of these points is

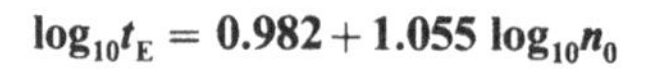

$$\log_{10} t_E = 0.982 + 1.055 \log_{10} n_0$$

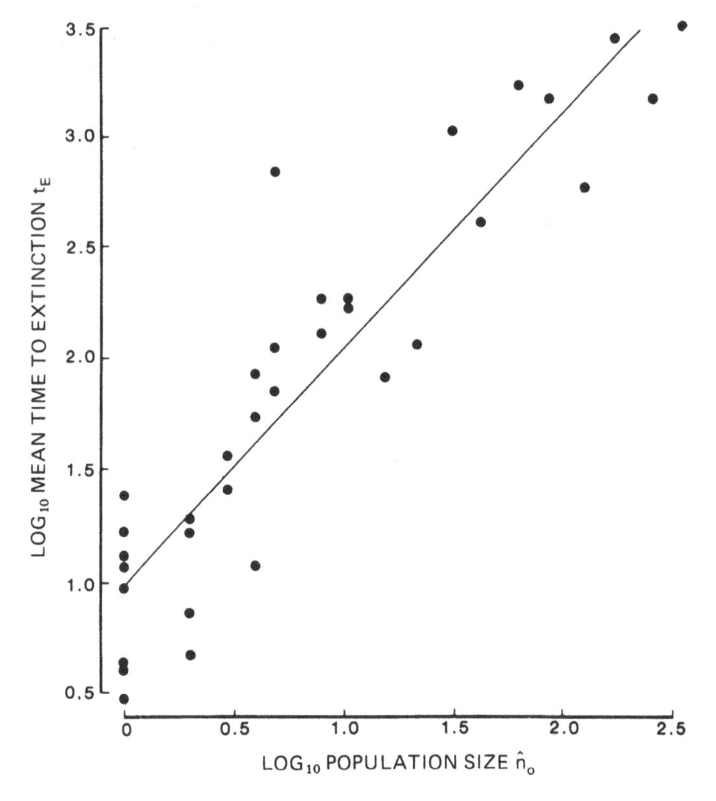

error. The sum of these two phases gives the time taken for the Ratchet to complete one turn.

For mutations of very small effect but moderately large frequency, the optimal class normally consists of a single individual; the establishment phase is much longer than the extinction phase, and the total time involved may be very large. These do not concern us very much, since their effect is so small that it will not lead to any appreciable loss of vitality. For mutations of large effect, the optimal class will comprise many individuals, the extinction phase will be much longer than the establishment phase, and again the total time involved may be very large. These do not concern us either, since in this case selection will effectively oppose the Ratchet. The most interesting category comprises mutations of small effect and moderate frequency, whose accumulation will lead to a gradual and eventually substantial loss of vitality. Either the extinction phase or the establishment phase or both may be substantial. Since we do not know how U varies with s, we cannot estimate the speed of the Ratchet directly. However, average values are very suggestive. Mukai (1964; Mukai *et al.* 1972) estimated the rate of mutation and the average effect of viability modifiers on the second chromosome of *Drosophila melanogaster*. The estimates of mutation rate per chromosome per generation in these two experiments were 0.141 and 0.172 respectively; since the second chromosome comprises about 40% of the *D. melanogaster* genome, these correspond to mutation rates of 0.35 and 0.43 per haploid genome per generation. The estimated average effects of these genes, as homozygotes, were 0.027 and 0.023 respectively. Since the mutation rate is underestimated and the average effect overestimated by Mukai's techniques, these figures will underestimate U/s and thus underestimate the speed of the Ratchet, so that the answers we shall get will be conservative.

The establishment phase is about 100 generations in both cases (94 and 126 generations for the two data sets). The extinction phase will be the time required for a cohort of $1.6n_0$ individuals to vanish, which will be approximated by $\log_{10}t_E = 1+\log_{10}n_0-\log_{10}(1.6)$. The total time required for the Ratchet to complete one turn is then the sum of establishment and extinction phases, as shown in figure 52 for the range of U/s between 10 and 20 which Mukai's results suggest as the most appropriate. In small populations the process is dominated by the establishment phase, and the Ratchet turns very quickly. In large populations the establishment phase is negligible, but the extinction phase may be very prolonged. For very many organisms, the Ratchet will be a potent source of decay. In a population of 10^{10} individuals, the Ratchet will turn once in about $10^{4.5}$ generations (with $U/s = 15$). For mutations with an average effect of

0.025 in a haplont, 100 turns of the Ratchet reduce the viability to less than 10% of its original value. If new mutations have an average dominance of 40% (the value obtained by Mukai), then their effect in heterozygotes will be 0.01, and 100 turns of the Ratchet will reduce viability to less than 40% of its original value in a diplont. Either figure would cause extinction for a protist reproducing by binary fission, which could not withstand reduction in viability of 50% or more, and would probably cause the demise of a multiparous organism. This would happen within $10^{6.5}$ generations, or between $10^{4.5}$ years (for a protist with many generations per year) and $10^{6.5}$ years (for an annual plant or animal). A population of 10^{10} individuals would therefore be ratchetted out of existence within a relatively short span of time.

10.4 Evading the Ratchet

The rate at which the Ratchet turns depends on *N*, *U* and *s*. It will turn very slowly if *N* or *s* are large, or *U* small.

Figure 52. The time required for the Ratchet to make one turn, in populations of different size. Based on the arguments in the text, and on Mukai's estimates of *U*/*s*.

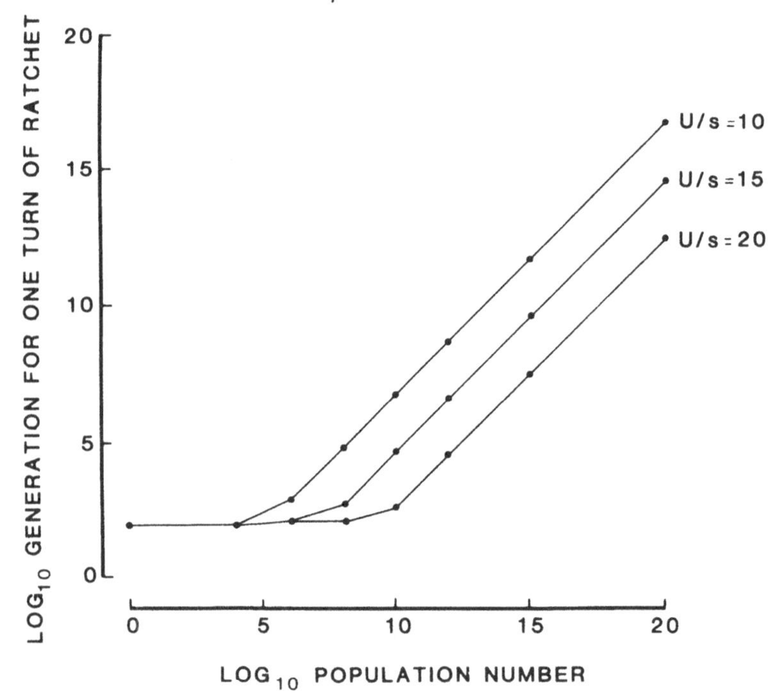

The population size N is defined ecologically; it is the group of organisms whose numbers are regulated in common. This might comprise all members of a single taxonomic species; at the other extreme, it might include different numbers from many species, if these compete with one another for the same resources. In many circumstances it will be difficult to decide how large N will be, even to an order of magnitude. However, there is no doubt that N will often be very large. Imagine a population of diatoms living in a lake. The population is regulated as a unit, perhaps through competition for limited supplies of phosphorus or silica, and is regulated quite separately from similar populations in nearby lakes. Population numbers may fluctuate enormously through the year – a difference of four orders of magnitude between spring and summer would not be unusual – making it difficult to give a single unambiguous value for N. One approach is to interpret a spring bloom as the 'progeny' which have been produced by the minimal population of the previous year, and from which the next minimal population will represent a sample of

Figure 53. Sex and size in algal unicells. The plotted points show the frequency of sexual forms in different size categories for all volvocacean unicells (○) and for *Chlamydomonas* alone (●). Solid line is least-squares regression of pooled data. From Bell (1985).

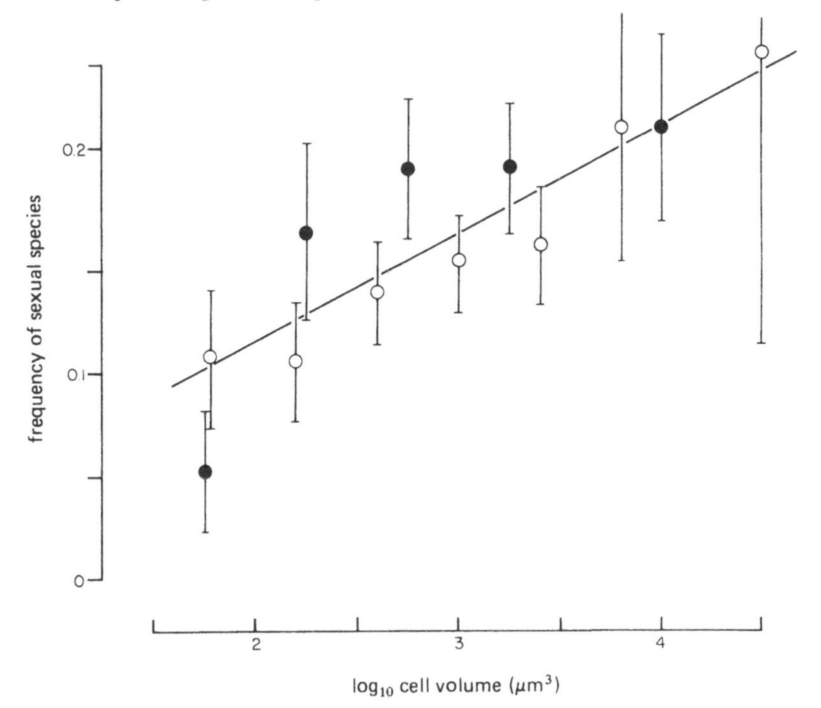

'adults'. In a population overwintering as resistant spores, the first generation of vegetative cells hatching from these spores in the spring would represent the adults. We might then interpret an organism with strongly seasonal dynamics as having a population size equal to the minimal number of vegetative cells and a generation time of one year. This minimal population is often of the order of 10^6 cells m^{-3}. A lake of 1000 m radius and with a photic zone extending to a depth of 10 m has an effective volume of the order of 10^6 m^3, and therefore supports a (minimal) population of some 10^{12} cells. This does not seem to me at all an excessive estimate of the population size of many planktonic protists, although of course it does rely entirely on the assumption that the population of the entire lake is regulated as a unit. By the same reasoning, many prokaryotes and protists will have populations of 10^{15} or more individuals. These values should be large enough to reduce the Ratchet to insignificance. With $N = 10^{12}$ and $U/s = 15$ the optimal class comprises some $10^{5.5}$ individuals, and 100 turns of the Ratchet will occupy more than 10^8 years. A very low rate of migration from neighbouring populations will be sufficient to halt the Ratchet entirely for all practical purposes. Recognizing many zones within the lake whose populations are regulated separately will not change this conclusion, since migration between them will be overwhelming.

Naturally, there will be very many protists whose population size never exceeds 10^{10}, and which will therefore suffer severely at the hands of the Ratchet. Among larger organisms the situation is clear: very few organisms with a mass of more than 1 g will maintain populations of as many as 10^{10} individuals. The evolution of large size therefore requires the prior evolution of sexuality; it is most unlikely that multicellular organisms could evolve from exclusively asexual lineages. The tendency for obligately asexual reproduction to be much more common among very small eukaryotes is well known (Bell 1982, p. 304). A striking example is shown in Figure 53; the frequency of forms known to be sexual is an increasing function of size in unicellular volvocacean algae, and even within the single genus *Chlamydomonas*. The larger, colonial members of the family all have sexual reproduction. The very general association between sex and large size can be interpreted in other ways (as in Bell 1982), but is certainly consistent with the view that very small eukaryotes are often sufficiently numerous to evade the Ratchet without the need for recombination.

Organisms with populations of fewer than 10^{10} individuals may still escape the Ratchet if U is sufficiently small. Whilst little can be done to reduce the rate of mutation per locus, the overall rate of mutation will be proportional to the number of mutable sites, and will therefore be low if the genome is small. Mitochondria provide a striking illustration of this possibility. Human cells each contain about 10^3 mitochondria, which constitute a closed population. With so small a population size, mitochondrial genomes should become degraded by the Ratchet very quickly. However, the mitochondrial genome is only about 1 % as large as the nuclear genome; instead of U/s in the range 10–20 for $s = 0.01$, U/s will be in the range 0.1–0.2. The optimal class will then represent a majority of the population, and most of the rest of the population will carry only a single mutation. The arguments developed above cannot be applied to this situation, since they assume that the frequency of unloaded individuals is very small. However, it is obvious that the Ratchet will turn very slowly indeed. For mutations which are nearly neutral, on the other hand, the small population size of germ-line mitochondria will mean that they will accumulate mutations rather rapidly, in a manner akin to ciliate macronuclei. Presumably that is why mitochondrial genomes are so variable, with almost all the variance occurring between individuals, rather than between mitochondria within the same cell.

In other organisms the situation is quite different. In yeast, for example, there are only about 20 mitochondria, each of which is so large that together they account for nearly 20 % of the yeast genome. With so small a population number and so large a genome, there is little doubt that yeast mitochondria would Ratchet very quickly. This is perhaps why mitochondrial genomes recombine rather freely in yeast. Recombination is also known to occur between chloroplast genomes in *Chlamydomonas reinhardti*, where they comprise about 15 % of the total genome. In this case, recombination has been detected only in the rare event of both gametes contributing a chloroplast to the zygote; I presume that it must also occur between the several genomes resident in the single chloroplast of normal cells, though this would be difficult to demonstrate.

The proposition that small genomes evade the Ratchet can be turned on its head: because of the Ratchet, large genomes cannot evolve in the absence of recombination. I have suggested above that large organisms cannot evolve in strictly asexual lineages, since their relatively small population size will be within the range where the Ratchet turns rapidly. Putting these two ideas together, the Ratchet will represent an insuperable obstacle to the evolution of large, complex organisms whose specification or operation requires a large genome. It is certainly a very striking fact that most of the history of life has been taken up with small unicellular forms: large multicellular organisms are an evolutionary novelty. The multicellular soma scarcely seems so radical an innovation that it should have taken 2 billion years to evolve. I believe that the reason for this extraordinarily long lapse of time may have been that large, complex organisms could not persist on a geological timescale in the absence of an efficient means of genetic recombination.

So far, I have assumed that the loci at which mutation is liable to occur have independent effects on fitness. It is quite possible, however, that loci interact, so that the fitness of the double mutant is less than $(1-s)^2$. The effect of this interaction would be to reduce the power of the Ratchet, since selection will more readily eliminate heavily loaded genomes. Suppose that the fitness of an individual with i mutations is

$$W_i = (1 - si^E)^i,$$

where E is a parameter which expresses the epistatic interaction between loci. If $E = 0$ then the loci have independent effects on fitness; for $E > 0$ the fitness of a multiple mutant is less than the product of the fitnesses of the single mutants. The properties of the frequency distribution of load at its deterministic equilibrium as a function of E are shown in Figure 54.

Figure 54. The effect of epistasis on the mean load and on the frequency of the unloaded class at equilibrium. The values plotted here were obtained by numerical iteration with $U = 0.14$ and $s = 0.01$, $U/s = 14$.

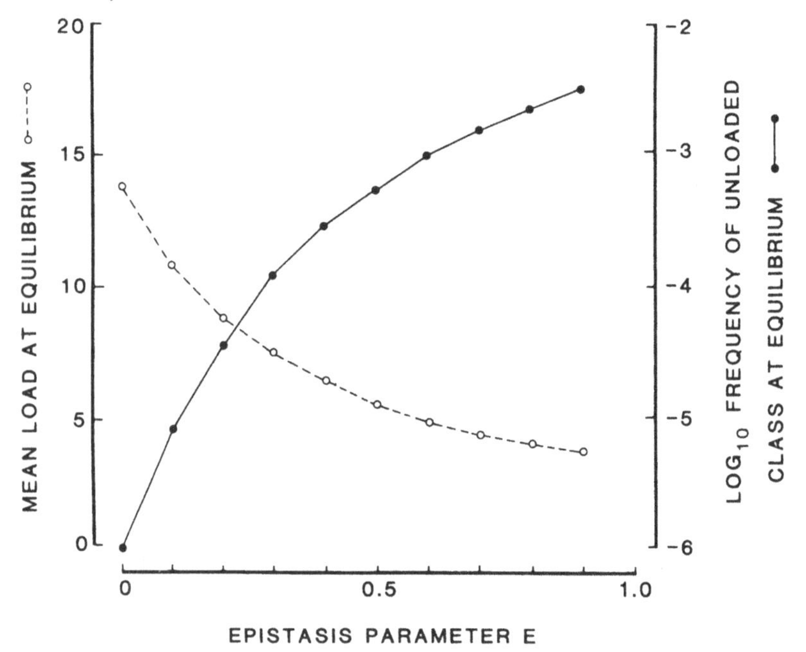

The mean load falls from U/s when $E = 0$ towards unity for large E. This implies that the frequency of the unloaded class becomes very great. Such extreme epistasis is implausible; more important is the fact that moderate degrees of epistasis produce rather large increases in n_0. As E increases from 0 to 0.1, for example, n_0 increases by nearly an order of magnitude from 10^{-6} to 10^{-5}, for $U/s = 14$. This will increase the extinction time proportionately. The Ratchet may therefore be rather ineffective in accumulating mutations at functionally related loci. Conversely, if there is negative epistasis, such that the fitness of the double mutant is greater than $(1-s)^2$, the Ratchet will turn more quickly.

To sum up, the Ratchet seems likely to be a negligible evolutionary force among prokaryotes and eukaryotic unicells which maintain population sizes of 10^{12} or more as an annual minimum. It will also be slowed down by the interdependence of functionally related loci, which will have an effect equivalent to that of reducing genome size. It will be a powerful force among organisms with populations of 10^{10} or fewer individuals, with genomes large enough to experience mutations which reduce viability by 1% at a rate of 0.5 or so per genome per generation. It will almost certainly prevent the evolution of large size and genetic complexity in the absence of efficient mechanisms of exogenous repair by recombination, and will doom asexual lineages of such organisms to extinction within a brief span of geological time.

10.5 How much recombination?

If exogenous repair through cross-fertilization is essential to prevent the deterioration of the germ line, exclusively parthenogenetic taxa, whether apomictic or automictic, cannot be very long-lived. This is almost certainly the case in metazoans, where the sporadic taxonomic distribution of parthenogenesis shows that it usually occurs in relatively short-lived branches from amphimictic stocks (see Bell 1982). The major exceptions are provided by bdelloid rotifers and chaetonotoid gastrotichs, both common and widely-distributed groups in fresh water. The bdelloids are apomicts in which neither males, gametes nor any form of sexuality have been reported; chaetonotoids are automictic or self-fertilized, with no known capacity for cross-fertilization. Both groups have amphimictic relatives, but their distinctive morphology and variety of form make it difficult to believe that they have arisen recently.

There are also several rather large groups of eukaryotic protists – cryptomonads, euglenids, trichomonads, amoebas, choanoflagellates and others – from which meiosis has never been described. Their existence raises some very awkward questions. For instance, it seems to be universally accepted that the ancestral eukaryote was a mitotic haplont, on the grounds that groups such as those I have mentioned above are primitive. If my arguments are sound, it seems questionable that such a creature could evade the Ratchet for the tens or hundreds of millions of years necessary to evolve meiosis and cross-fertilization. Either the earliest eukaryotes were always extremely abundant; or they possessed some alternative open genetic system; or meiosis and cross-fertilization were ancestral to mitotic replication. More generally, the present existence of many apparently perennially mitotic protists runs counter to the near-universality that I have claimed for the Ratchet. If some form of genetic exchange is indispensable, nothing is known of how frequent such exchange must be. It is quite plausible that very low rates of recombination might be sufficient to halt the Ratchet in large populations. There is clearly a need for further theory; I have made only some rough calculations.

Since the optimal class becomes extinct after about $10n_0$ generations in the absence of recombination, it will be completely reconstituted in sexual populations if on average 0.1 unloaded genomes are produced per generation by recombination. The rate of recombination needed to halt the Ratchet is, then, simply the rate which will supply on average one unloaded genome every ten generations.

This rate is not a function of population size: one unloaded genome produced every ten generations will halt the Ratchet in a population of any size. A very low rate of recombination in a very large population is therefore just as effective as a very high rate in a much smaller population. If the population is too large, then, as we have already seen the Ratchet will not turn anyway. Conversely, if the population is very small, no amount of recombination will be sufficient to generate enough unloaded genomes. We can therefore identify three regimes. In very small populations, recombination will be ineffective, whilst in very large populations it will be unnecessary; it is in populations of intermediate size that recombination will be relevant.

To construct a theory, suppose that the genome comprises a single large chromosome, along which any number of crossovers r may occur. We wish to calculate the value of r which is just sufficient to create an additional 0.1 unloaded gametes per generation. These are needed to balance the stochastic loss of the unloaded class; they must therefore be supplied from the loaded gametes. To calculate the rate at which unloaded gametophytes are produced from zygotes formed by the fusion of two loaded gametes is quite difficult; fortunately, we can argue that at equilibrium the rate of gain of unloaded genomes from the loaded classes must be equal to the rate of loss of unloaded genomes from the unloaded class. This sets us the easier task of calculating the rate at which unloaded genomes are lost during gamete fusion and meiosis.

Consider an unloaded gamete. When this fuses with a gamete bearing i mutations ($i > 0$), and j crossovers occur, let the frequency of unloaded genomes among the gametophytes produced by meiosis be X_{ij}. Since there are four meiotic products, the number of unloaded genomes produced is $4X_{ij}$, and the number of loaded genomes produced is likewise $4(1-X_{ij})$. Only two of these gametophytes are expected to survive if the population is to remain stationary in number; after the random loss of two gametophytes, therefore, the number of loaded genomes produced is $2(1-X_{ij})$. Since one loaded gamete entered into fusion, the extra number of loaded genomes produced is $2(1-X_{ij})-1 = 1-2X_{ij}$. We have to cumulate this number for all the crossovers permitted, and weight the result by the frequency of gametes bearing i mutations. If there is no recombination, then $X_{10} = \frac{1}{2}$, and $1-2X_{10} = 0$; we need consider only those zygotes in which the partner of the unloaded gamete bears $i > 0$ mutations, and in which $j > 0$ crossovers occur. The total number of loaded genomes created from the unloaded class of n_0 gametes is thus

$$n_0 \sum_{i=1} \left[P_i \sum_{j=1} (1-2X_{ij})Pr(j) \right]$$

where $Pr(j)$ specifies the probability that j crossovers occur. This number must exceed some critical value, say K, if the Ratchet is to be halted; we know that $K \approx 0.1$.

Suppose that there is only one crossover. This is almost certain to combine a sequence with no mutations, from one of the two chromatids produced by the unloaded gamete, with a mutated sequence from the loaded gamete. The other unloaded chromatid is transmitted intact; thus $(1-2X_{ij}) = \frac{1}{2}$. If a second crossover occurs, it is equally likely to involve either chromatid from the unloaded gamete, the one which has already undergone crossing-over or the uncontaminated one. The frequency of unloaded gametophytes therefore drops to $\frac{1}{8}$, and $(1-2X_{ij}) = \frac{3}{4}$. Extending this argument, it is obvious that if there are j crossovers, the frequency of unloaded gametophytes produced by meiosis will be nearly $2^{-(j+1)}$ when the number of mutations i borne by the loaded gamete is fairly large. The real frequency will be somewhat greater – if $i = 1$, for example, $X_{ij} = \frac{1}{2}$ irrespective of j – and therefore

$$n_0 \sum_{i=1} \left[P_i \sum_{j=1} (1-2^{-j})Pr(j) \right]$$

will provide a conservative estimate of the rate at which unloaded genomes are being produced. It will nevertheless be very close to the correct value if U/s is fairly large, since in that case the lightly-loaded gametes which introduce an error into the approximation are relatively rare. If we suppose that there is a fixed small probability that a crossover will form in each of the small segments into which the chromosome may be divided, the number of chiasmata per bivalent will follow a Poisson distribution. This is a biologically reasonable model, though in practice interference between chiasmata will often cause deviations from Poisson expectations. The parameter of the Poisson series is r, the mean number of crossovers per chromosome. Summing the series in the expression above shows that the critical value of recombination which just prevents the Ratchet from turning is

$$r^* = -2 \log_e [1 - K/\hat{n}_0 (1 - \hat{P}_0)]$$
$$\approx -2 \log_e (1 - K/\hat{n}_0),$$

assuming that the deterministic frequency of the optimal class $\hat{P}_0$ is the same in populations with and without recombination, which will be nearly true if the population is reasonably large. The model of chiasma distribution has little effect on this conclusion. If there is no variance in chiasma frequency, exactly r crossovers being formed on each chromosome, the critical value of r is

$$r^* \approx -2 \log_2 (1 - K/\hat{n}_0) = -1.44 \log_e (1 - K/\hat{n}_0).$$

Greater variance than Poisson is created by a geometric distribution with parameter r', the mean number of crossovers being $r = r'/(1 - r')$ and the critical value

$$r^* \approx 2(K/\hat{n}_0)(1 - K/\hat{n}_0),$$

which approximates the Poisson solution for $\hat{n}_0 > 1$.

The Poisson solution shows that no amount of recombination will halt the Ratchet if $\hat{n}_0 < K$. Clearly, if the number of unloaded genomes K supplied by recombination exceeded the number present $\hat{n}_0$ the population would not be in equilibrium. Consequently, populations with high mutation rate or small size will deteriorate whether or not recombination occurs. If the *Drosophila* estimate of $U/s \approx 15$ is representative, eukaryote species with fewer than about $N = 10^5$ members will be short-lived in evolutionary time, even if they are obligately sexual. Among more abundant organisms with $\hat{n}_0 > 1$, $(1 - K/\hat{n}_0)$ can be approximated as

$\exp(-K/\hat{n}_0)$, and with $K \approx 0.1$ there is a very simple relationship between n_0 and the critical value of recombination:

$$\log_{10} r^* \approx -1.6 - \log_{10} \hat{n}_0.$$

One way of appreciating the implications of this result is to say that, ignoring any correlation between mutation rate and population size, the product of the recombination rate (as mean number of crossovers per genome) and population size will be a constant, with a numerical value of about 10^6. Very low rates of recombination will be adequate to halt the Ratchet in large populations; for $N = 10^{10}$, $r = 10^{-4}$ is sufficient, so sexual episodes need be only infrequent in small heterogonic metazoans such as *Volvox*, rotifers or cladocerans. In protists or prokaryotes with populations of 10^{15} or so individuals, the critical rate of recombination is too low to be detectable except in very large experiments. Conversely, small populations require high rates of recombination, and more or less obligate sexuality may be a precondition for the long-term survival of species with fewer than about 10^7 members.

10.6 The necessity for outcrossing

The Ratchet sets a limit to the longevity of many asexual germ lines, which will almost certainly not allow them to persist through geological time. This limit represents the limit of effectiveness of endogenous repair systems; to extend the lifespan of the germ line, endogenous repair must be supplemented by an exogenous system involving variation and selection.

The relevant variation is variation in the number of defective genetic elements per genome. If the genome bears a single deleterious mutation, mitosis will replicate the defect along with the rest of the genome, and it will continue to be borne by all subsequent members of that line of descent. However, a process which causes variation may create some offspring which lack the mutation, and will have thus recovered an uncontaminated genome. In closed genetic systems, such as self-fertilization, variation is created purely through new combinations of elements already present in the genome. Thus, half the offspring of an individual which has recently acquired a mutation will themselves bear a single copy of the mutation, but the other half will be homozygous, a quarter bearing two mutations and a quarter none. In open genetic systems, such as cross-fertilization, variation is created through the recombination of material arising in different lines of descent. If two individuals suffer a single mutation, it will almost certainly occur at different loci; half their offspring will bear one or the other mutant allele, a quarter will be heterozygous for both mutations and a quarter will be free of mutation. Both open and closed systems, therefore, create variance among progeny with respect to mutational load. Nevertheless, only open systems are effective in exogenous repair.

The reason why closed systems cannot do much to slow down the Ratchet is sketched in Figure 55. First, imagine a haploid apomict, which to begin with has no mutations. After a period of time which depends on

Figure 55. The fate of new mutations in closed genetic systems. (*a*) in a haploid apomict, the rate of mutation at a given locus is m; thus any individual gives rise to wild-type offspring with probability $1-m$ and a mutant offspring with probability m. (*b*) in a diploid automict, a locus with a rate of mutation m for each allele will become heterozygous for a mutation with probability $2m$ (approximately, if $m \ll 1$). The heterozygote will segregate into two homozygous lines, one of which carries two copies of the mutation. (*c*) a N-ploid amitotic nucleus will acquire a mutation at a given locus with probability Nm; $1/N$ of the lines descending from the single-mutant individual will bear N copies of the mutant allele, while the remaining lines will be homozygous wild-type.

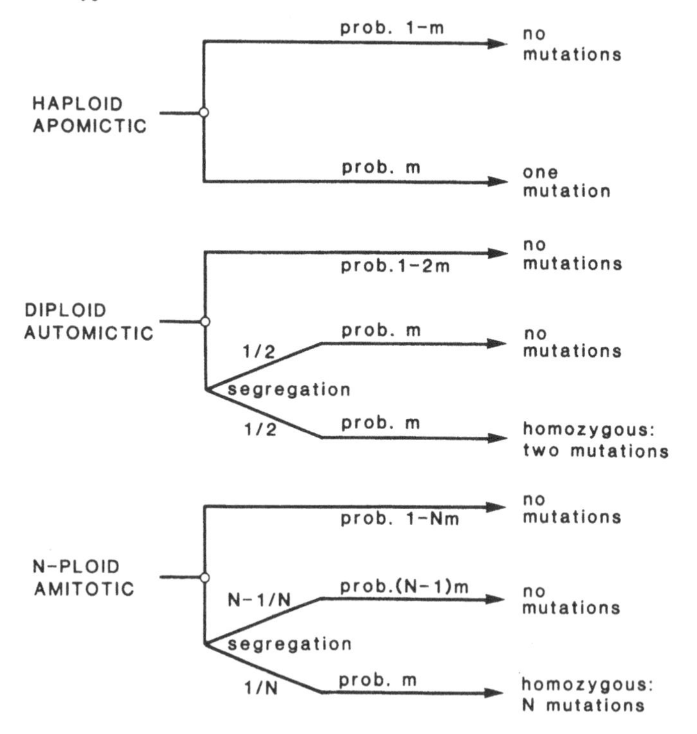

the rate of mutation per genome U, a mutation occurs. The population has now split into two classes, an unloaded class with frequency m and a class with a single mutation whose frequency is $1-m$. Clearly, selfing will not in itself make any difference to this process, since the gametophytes produced by a haploid selfer are identical with their parent. It is at first sight a little surprising that the same is true for a diploid selfer. Once a deleterious mutation has appeared, it will segregate rather quickly: within a short span of time, relative to the time taken to acquire a particular mutation, almost all the descendants of the original mutant will be homozygous either for the wild-type or for the mutant allele. The frequency of the homozygous mutant class will now be half the corresponding frequency of the hemizygous mutant class in the haploid apomict. But because the selfer is diploid, it will suffer twice as many mutations per generation. The time taken to create a homozygous mutant line in the selfing diplont is therefore nearly the same as the time taken to acquire a mutant line in the haploid apomict. Once homozygous, no further variance in copy number can be generated by selfing. For alleles which have the same effect in homozygous as in hemizygous state, therefore, the Ratchet moves at the same pace in a diplont with selfing as in an entirely asexual haplont. A rigorous proof of this conclusion is given by Heller and Maynard Smith (1979), who also show that the degree of dominance of new mutations of small effect has little influence on the rate of the process. A quite different point of view has been advanced by Shields (1982), who argues that inbreeding (short of complete self-fertilization) halts the Ratchet because of the effectiveness of selection in removing deleterious recessive mutations once these have been made homozygous, while a restricted amount of outbreeding will be sufficient to restore the optimal class.

Similar arguments apply to segregational systems which create variance, such as amitosis. If a particular element is present in n copies, one of which mutates, then $1/n$ of the descendant lines will bear the mutation in all n copies. The rate at which fixed lines are formed, per mutation, is thus $1/n$ times the rate at which they arise in an asexual haplont; but since the amitotic nucleus is n-ploid the mutation rate is n times as great, and the overall rate of formation of mutated lines is the same in a high-ploid amitotic organism as in a haploid apomict. It follows that the Ratchet will operate with full power against ciliate macronuclei, and indeed against cytoplasmic elements generally.

A diploid apomict will survive for longer than a haplont, and in general higher ploidy will purchase greater longevity. This is simply because the full effect of partly recessive genes on viability will take longer to be manifested when there are several or many copies per nucleus. We can imagine a genome, initially uncorrupted, which begins to acquire mutations. This initial period will be longer for lines of greater ploidy. Eventually a great many loci will bear mutations, almost all of them heterozygous. Greater ploidy then gives no protection; for instance, a diploid apomict in which every locus is heterozygous for completely recessive mutations will degenerate as quickly as an uncorrupted haplont. The initial period, during which mutations accumulate without any crippling loss of function, may be quite extensive; it will be proportional to ploidy and to the inverse of the mutation rate. It is a very striking fact that most apomictic metazoans are polyploid; indeed almost all polyploid metazoans are apomictic (see Bell 1982). Polyploidy is also very common among self-fertilized angiosperms. It may be that most asexual lineages are indeed of very short duration, so that we observe mostly the longer-lived lines in which apomixis is combined with high ploidy. Nevertheless, once this initial period is over, and deleterious recessive mutations have occurred at a substantial fraction of the loci, the Ratchet will cause deterioration in vitality as inevitably and nearly as swiftly as it does in haplonts. In particular, it should be noted that the calculations I have made of the speed at which the Ratchet turns, on the basis of Mukai's estimates of the mutation rates of viability modifiers, assume that these estimates apply to a haplont. Since *Drosophila* is in fact diploid, it may be more realistic to assume that $U \approx 1$ rather than $U \approx \frac{1}{2}$; since this implies that $U/s \approx 30$, my estimates of the speed at which the Ratchet turns and the amount of recombination required to halt it may be very conservative and my case for the importance of the Ratchet correspondingly understated. For further discussion of the relationship between mutation, polyploidy and the Ratchet see Manning (1983; also Manning and Dixon 1986).

In short, the variance generated by different combinations of genes from the same nucleus has little effect on the operation of the Ratchet. Any organism with a closed sexual system will decay through time at more or less the same rate as a strictly asexual lineage. We may then propose, as a general rule, that no germ line can persist for geologically substantial periods of time in the absence of cross-fertilization. The exceptions will be very numerous organisms with small genomes.

A similar rule is of some antiquity: Charles Darwin, and Andrew Knight before him, believed that no organism may self-fertilize in perpetuity. This belief, however, was founded on the rapid and general decline in vigour due to inbreeding depression, and therefore attributable to the sudden uncovering rather than the gradual accumulation of deleterious genes of low penetrance.

The population genetics of vegetative reproduction have not been worked out, but I see no reason to suppose that reproducing by fission will grant any general dispensation from senescence. Mutations will occur in stem cells, and when these proliferate during embryogenesis each mutation will be cloned. The embryo will then be a mosaic for all the mutations which have arisen in its (stem-cell) line of descent. In the absence of meiosis, it is difficult to see how this steadily increasing burden of somatic mutations can be shed.

The Ratchet operates in closed sexual systems because any locus that has become homozygous for a mutation cannot subsequently recover its original state. The consequence is that a diploid automict becomes Ratchetted at about the same rate as a haploid apomict. However, although the Ratchet acts on homozygous loci in an automictic diplont, it does not act on heterozygous loci, since they can give rise by assortment to wild-type homozygotes. Whether or not this will restore vitality in isolate cultures depends on the dominance of the mutation. Imagine, for instance, that a particular mutation is almost completely recessive, but severely depresses viability in homozygotes. If an automixis occurs after such a mutation has arisen, the population will be split into two lines, one homozygous for the mutation and the other lacking it entirely. Because of its severe effect on homozygote viability, the investigator is almost certain to choose the uncorrupted line to continue the culture. Provided that the mutation is not completely recessive, an apomictic line will gradually deteriorate as mutations build up in homozygous state, while an automictic line is nearly certain to be purged by each successive automixis. This is made feasible by dropping the assumption made in previous models, that mutations have independent effects on fitness, which for mutations of small effect amounts to codominance. If more than one locus has mutated before automixis intervenes, the frequency of uncorrupted recombinants will be correspondingly lower, and many of the individuals chosen will bear one or several mutations. This frequency will be increased if there is interaction between nonallelic as well as between allelic genes, so that the viability of multiple homozygotes is less than the product of the viabilities of the homozygotes taken separately. The accelerated decline of viability observed in isolate cultures is possibly evidence for such epistatic effects, which were also observed by Mukai for his viability modifiers in *Drosophila*. Moreover, the investigator chooses four or five individuals to restock all the lines of a series, and may subsequently transfer individuals from a flourishing line to one that has become extinct. Only one of these individuals need lack mutations for the Ratchet to be halted.

Thus, if mutations at the same and at different loci are not completely recessive but have markedly non-multiplicative effects on fitness, automixis is capable of restoring complete or nearly complete vigour, as the analysis of Woodruff's cultures showed to be the case. This will occur only if the number of asexual generations between successive automixes is rather small; in Woodruff's cultures it was less than fifty. As it becomes greater, the frequency of uncorrupted recombinants becomes very small even with large amounts of dominance and epistasis, and the automixis will merely cause massive mortality, as we have also found to be the case.

Isolate cultures therefore represent a special case of the general proposition that the mean fitness of sexual populations, whether selfed or crossed, will exceed that of asexual populations when deleterious mutations have synergistic effects on vitality. Why this should be so can be understood by considering an extreme case in which individuals with k or fewer mutations are unaffected while those with $k+1$ or more mutations immediately die. An asexual population will accumulate such mutations until all individuals bear exactly k mutations; naturally, all those which subsequently undergo mutation so as to bear $k+1$ mutations die and are thus removed from the population. The mean load of an asexual population in such a model is therefore k. In a sexual population, recombination will create a frequency distribution of mutations per genome. Even if all individuals at some point were to bear k mutations, some of their progeny will bear fewer than k, while others will bear $k+1$, $k+2$, $k+3$ and so forth. Since all those which bear $k+1$ or more will die, one death will often rid the population of more than one mutation. Since a sizeable fraction of the surviving offspring are thus enabled to bear fewer than k mutations, the mean load of the sexual population will be less than k at equilibrium. In the more general case, we can imagine that the logarithm of viability might decline more steeply than linearly with the number of mutations per genome. In deterministic populations, this implies that recombination will have greater mean fitness than populations with none. Kondrashov (1984) has recently analysed this situation. In finite populations, it implies that each successive turn of the Ratchet will take longer to complete, since the fitness of the optimal class, relative to the mean fitness of the population, will increase with the mean load in the population.

10.7 The breakdown of repair systems

The analogy of the craftsman and the chairs which I used above is incomplete in one major respect. It is clear that errors in the chairs will tend to accumulate, even if the skill of the craftsman remains undiminished. But we have no warrant for assuming that craftsmen will not themselves deteriorate, just as their chairs do. This does not in itself introduce any new element into the situation: errors will simply accumulate at some steady rate in both craftsmen and chairs. But suppose that each craftsman is mortal, and besides making a new generation of chairs must also instruct a new generation of craftsmen. Because these instructions are themselves subject to error, the ability of successive generations of craftsmen to detect and remedy errors in chairs must fall. Even worse, their ability to transmit information about error-detection and repair must also fall. In the chairs themselves, errors will not only accumulate, but will accumulate at an ever-increasing rate.

The construction of self-replicating machines has been discussed by von Neumann (1966) and Spiegelman (1971). The von Neumann machine uses a pile of raw material to construct a replica of itself, under the control of a tape bearing a linear sequence of instructions. It was imagined that such a process could be repeated indefinitely, the result being an immortal line of self-replicating automata. Whether or not this conclusion is true depends on whether or not the tape is itself copied, and, if so, whether the copying device is itself part of the machine. If the same individual tape is loaded into each new machine by some external operator, then the line of machines could be perpetuated for as long as desired. Let us suppose instead that the tape is itself copied in each generation, by a second machine. Because no copying device is perfect, this second machine will occasionally introduce an error into the instruction tape. Because there is no way in which errors once introduced can be subsequently detected and corrected, these errors will tend to accumulate. With a population size of one, the Ratchet guarantees a swift and irreversible decline in performance. The immortality of the von Neumann machine thus depends on the existence of an immortal tape.

Moreover, the machine also requires an external operator, responsible for reloading, and if necessary copying and correcting, the instruction tape. An autonomous self-replicator would need one additional feature: a final instruction on the tape, to copy the tape itself (including the final instruction) and then load it into the new machine. This is the machine proposed by Spiegelman (1971). However, any such machine will degenerate more rapidly than a von Neumann machine with an external tape copier. Errors will accumulate in the tape, as before; but they will accumulate in particular among the copying and error-correcting instructions. These errors will decrease the fidelity with which all instructions are subsequently copied – including the copying and error-correcting instructions themselves.

It is the element of self-reference in this process which makes the problem such a deep one. It makes it impossible, even in principle, to construct a self-replicating machine that is both finite and immortal. In living organisms, heredity takes the form of a physically-embodied, coded message passed on through the germ line. The information contained in this message will be continually degraded by random events, and a repair system is therefore essential to the perpetuation of the line. To be transmitted, this repair system must itself be encoded. But it will itself become degraded through time, so that its ability to repair itself will be lost. It is no use appealing to a secondary repair system whose function is to repair the primary system, since any secondary system will itself require repair. As the precision of the repair system is lost, not only will the soma deteriorate, but the repair system will be unable to specify itself correctly. The probability that a new error will become incorporated into the repair system is then directly related to the current level of error. It is this positive feedback which ensures that errors will accumulate at an ever-increasing rate.

A simple model will clarify the quantitative implications of this idea. There is no selection, and we can imagine that it refers to an organism which gives birth to a single offspring before dying. Suppose that there are R mutable sites coding for the repair system, some fraction F of which are mis-specified in a way which decreases the precision of repair. The initial probability of mis-specification, in an uncorrupted genome, is q_0 per site per replication, and this probability is increased by a factor $(1+k)$ by each additional error in the repair system itself. The number of mis-specified sites at any given time will then be equal to the number mis-specified in the previous generation, plus the product of the number of sites which were previously specified correctly and the prevailing error rate per site:

$$RF_{t+1} = RF_t + (R - RF_t)\, q_t$$

where $q_t = q_0(1+k)^{RF_t}$. The increase in the number of sites which are mis-specified is then

$$\Delta F_t = (1 - F_t)\, q_0 (1+k)^{RF_t}.$$

If F_t and k are reasonably small, this is approximately equivalent to $\mathrm{d}F/\mathrm{d}t = q_0 \exp(kRF)$. The time required to pass from any initial error frequency F_0 to any final value F_t can then be got by integration. If F_0 is small, this time is

$$t = (1 - e^{kRF})/kRq_0.$$

We can now interpret F as the amount of damage that causes a complete breakdown of function, and ask how long this will take to accumulate. First, we can see that k and R will affect t only through their product kR. Moreover, they are probably inversely related: the effect of a unit change is likely to be smaller when the total number of sites is larger. I shall guess that kR is of order unity. Then if F_t is fairly small, the time taken for a catastrophic loss of function is roughly equal to the reciprocal of the original error rate. Indeed, the form of the equation for t makes it clear that q_0, being a small number, will dominate its value, so despite the approximate nature of the argument it is unlikely that $t \approx 1/q_0$ will be far off. Now, even with an uncorrupted repair system the probability of error is unlikely to be less than 10^{-6}–10^{-7}; the number of replications before catastrophe is then 10^6–10^7, a timescale similar to that of the Ratchet in large populations.

When there is only a single genome in each generation, there can be no selection; these arguments cannot therefore be extrapolated directly to populations of many individuals in which the accumulation of mutations is opposed by selection. We can think about the dynamics of such populations by distinguishing between two types of element in the genome. One type includes loci which specify metabolic or structural proteins, which I shall call collectively the vegetative genome. All the arguments in previous sections have been concerned with genes of this sort. The second type includes loci which are responsible for DNA replication and proofreading, which I shall call collectively the repair genome. The greater the load in the repair genome, the greater will be the rate of (uncorrected) mutation in the vegetative genome; but the rate of mutation in the repair genome itself will also be greater, creating positive feedback which tends to increase the rate of mutation without limit. The first question is, then, whether or not there is any stable distribution of load, either in the vegetative or in the repair genome, when mutation is opposed by selection against heavily-loaded individuals. One might easily imagine that the positive feedback of mutation rate within the repair genome might fuel a runaway process in which the mean load increased without limit.

I have investigated this possibility by simulating a population in which the vegetative and repair genomes each comprise a fixed number of loci,

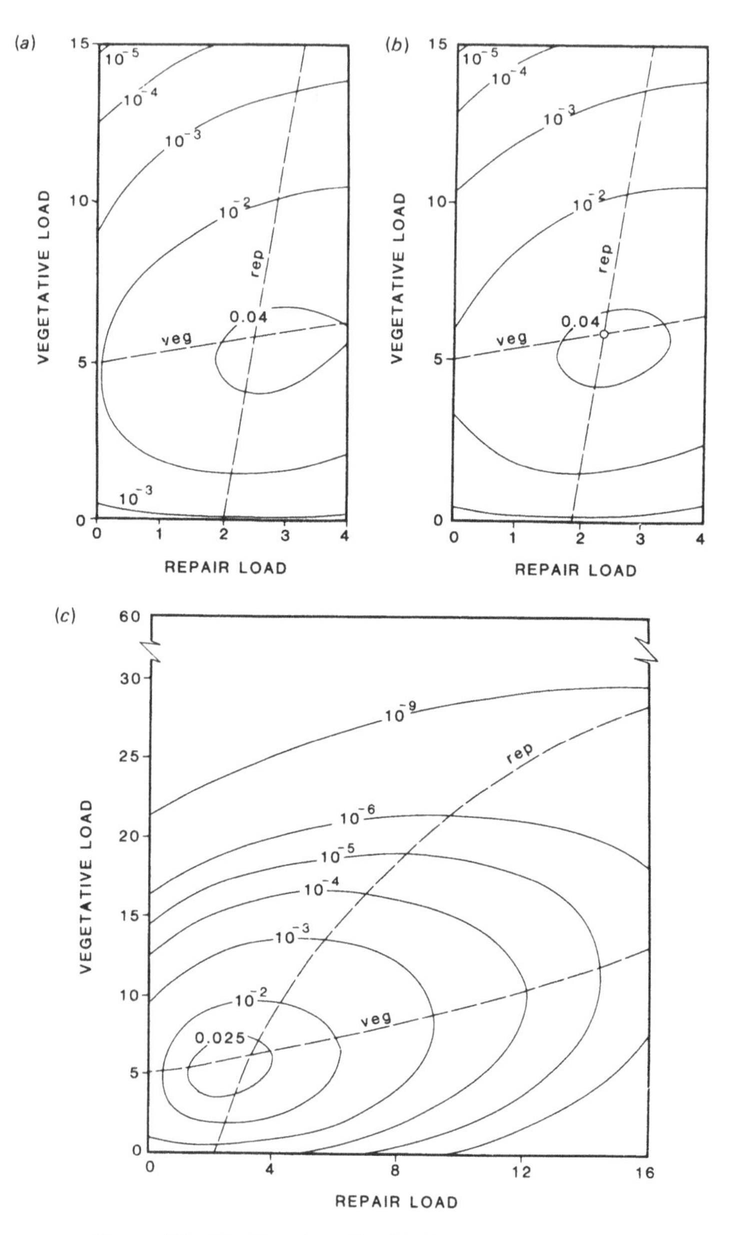

Figure 56. The bivariate distribution of load in the vegetative and repair genomes at equilibrium. The contours are frequencies after 500 generations in a model with 4 repair loci and 15 vegetative loci in a haplont. The broken lines are regressions: they represent the mean vegetative load for a given repair load (veg) and the mean repair load for a given vegetative load (rep). (*a*) An exact model. Among individuals whose repair genome was uncorrupted, the mean number of mutations

the number of mutations occurring in the genomes of each individual being a Poisson-distributed random variable whose mean is a function of the number of mutations currently borne in the repair genome. When both genomes are small this model showed that the load does not increase without limit; instead, the bivariate distribution of the repair and vegetative loads approaches a stable state after a few hundred generations (Figure 56(*a*)). This result might conceivably depend on the rather modest upper limit on the total number of mutations which can occur. To simulate larger genomes, I used an approximate model in which the vegetative and repair genomes may each acquire a single mutation per generation, but never acquire more than one; provided that the rate of mutation per genome is small this produces results very similar to those of the previous, more exact model (Figure 56(*b*)). When much larger genomes are simulated using this second model, a stable bivariate load distribution closely resembling that for smaller genomes develops within a similar timespan (Figure 56(*c*)). We can therefore conclude that positive feedback of mutation rates within the repair genome does not (necessarily) lead to a runaway increase in load but is instead effectively opposed by selection so as to establish a stable joint frequency distribution of vegetative and repair load in an infinite population. It is perhaps worth mentioning that the vegetative load, among individuals with any given repair load, has a Poisson distribution.

There are hints that a similar sort of process might operate in some ciliates. Some strains of *Tetrahymena pyriformis* have defective (hypo-diploid) micronuclei, and cannot accomplish reciprocal fertilization when mated with a normal cell. Instead, the migratory nucleus contributed by the normal partner doubles up its genome within the defective cell. This new homozygous diploid micronucleus gives rise to a macronuclear

arising per generation in the vegetative genome was 0.05, with Poisson variance; the corresponding rate for the repair genome was $(4/15) \times 0.05 = 0.0133$. An individual whose repair genome bore r mutations had a mutation rate in both the vegetative and repair genomes $(1+k)^r$ times as great as that of an individual whose repair genome was intact, with $k = 0.1$. The selection coefficient was 0.01. (*b*) An approximate model. This is identical to (*a*), except that any given individual acquired no mutation with probability 0.95 and one mutation with probability 0.05 per generation; no more than one mutation was permitted. Note the close similarity of (*a*) and (*b*), showing the adequacy of the approximation. (*c*) This figure uses the approximation developed in (*b*) to calculate the distribution of load in a much larger genome, with 16 repair loci and 60 vegetative loci. Note that enlarging the genome by a factor of 4 does not alter the result substantially.

anlage as usual, but this fails to develop, and the cell retains the old macronucleus. The result of this odd process – known as 'genomic exclusion' – is a cell with a new micronucleus but an old macronucleus. Cells which have been formed in this way are rejuvenated, and normal levels of fission rate and exconjugant survival are restored to senescent clones. The micronucleus is thus involved with vegetative senescence in *Tetrahymena*, despite the lack of transcription from the micronuclear genome, to an extent that is almost certainly not true in *Paramecium*. Even more interestingly, the rejuvenescence wears off within 50–100 generations, with vitality declining and overt symptoms of damage appearing in the micronucleus (Weindruch and Doerder, 1974). A reasonable interpretation of these facts is that the micronucleus has some vegetative function, detectable through its effect on senescence, while the macronucleus contributes to the repair of the micronuclear genome. We then have an approach to a physical realization of the vegetative and repair Ratchets. In *Paramecium*, Smith-Sonneborn (1971) found that the effect of treatment with ultraviolet light on the fission rate of cells subsequently kept in the dark increased with clonal age. This damage can be made good by exposure to light, until an age of about 150 fissions, beyond which it is irreversible. Similar observations have been made on *Spathidium* by Williams and Williams (1965). They suggest an accumulation of errors in the repair system.

The repair load is kept within bounds because genomes with greater load are less fit. This fitness is not an intrinsic property of the repair load; considered by itself, a mutation in the repair genome is selectively neutral. However, genomes with many mutations at repair loci are likely to acquire many vegetative mutations also, and it is the resultant deficiency of vegetative function which is counterselected. Selection against mutations in the repair genome is therefore an indirect response to selection against vegetative mutations, and will occur only to the extent that there is a positive genetic correlation between vegetative and repair loads. This is shown in Figure 56 by the line which plots the mean vegetative load for genotypes with a given repair load. This correlation is maintained in asexual populations, such as those simulated, because any mutation in the repair genome is indissolubly linked to any mutations which it causes in the vegetative genome. The only effect of recombination will be to reduce the correlation between the repair and vegetative loads (Figure 57). The indirect response of the repair load to selection against vegetative mutations will therefore be weaker when recombination occurs. It follows that in deterministic populations, the mutational load in both repair and vegetative genomes will be greater when reproduction is sexual than when it is asexual.

The greater repair load, and thus the greater vegetative load, borne by sexual populations in stable environments does not, of course, imply an irreversible deterioration in performance. In finite populations, one can imagine that the genome is subject simultaneously to two Ratchets, a repair Ratchet and a vegetative Ratchet. The optimal class will be smaller in a sexual population than in a comparable asexual population, as a consequence of recombination between the repair and vegetative genomes. Nevertheless, recombination within the repair genome will still recreate

Figure 57. The correlation between the vegetative and repair loads in sexual and asexual populations. The values plotted were attained after 500 generations in the model shown in Figure 56(*b*). For the sexual population the model is identical except that zygotes are formed by random gamete fusion; during meiosis there is free recombination between the vegetative and repair genomes, but no recombination within either.

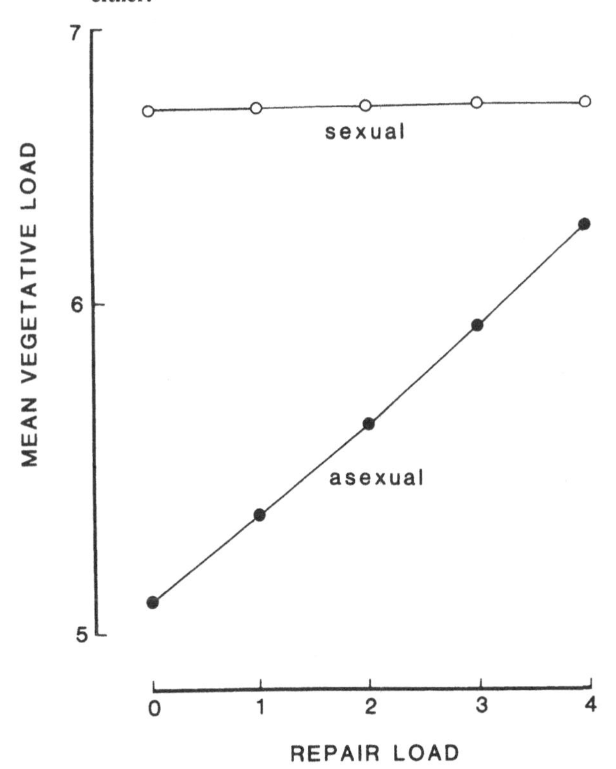

unloaded repair genomes in the sexual population, while the asexual population will be unable to recover the unloaded class once it has been lost. There will therefore be an irreversible tendency for mutation rates to increase, and moreover to increase at an increasing rate, in the asexual population. As the repair Ratchet turns, the mutation rate increases, and the vegetative Ratchet therefore turns more quickly, hastening the decline and eventual breakdown of performance. To see how important an effect this might be, we can calculate the time taken for the repair Ratchet to turn 100 times, when each turn increases the effective mutation rate by a factor of $(1+k)$. If there were no repair Ratchet then $k = 0$, and we can calculate the time involved through knowing, from the results given above, how the time taken for each turn varies with the size of the least-loaded class. Now, suppose that $k > 0$. The time taken for the repair Ratchet to turn for the first time will be the same as if $k = 0$, assuming the initial mutation rates to be the same. We can then compare the time taken for the next 100 turns with the corresponding time when there is no repair Ratchet. The ratio of these two numbers is

$$\frac{\exp(U/s)}{100} \sum_{i-1}^{100} \exp[-(1+k)^i(U/s)]$$

which will vary between zero (if k is very large) and unity (when k is zero). It is mapped as a function of k and U/s in Figure 58.

What this figure shows is that the autocatalytic breakdown of the repair systems seems to be of little consequence if k and U/s are both small, and will be substantial only when both are large. The selection coefficient for repair mutations will be smaller than that for vegetative mutations, in proportion to the regression of the repair load on the vegetative load. However, this effect is likely to be swamped by the much lower overall rate of mutation in the repair genome, assuming it to be much smaller than the vegetative genome. We do not know the size of the repair genome for any eukaryote; but if it is as small as 1 % of the total genome then U/s for the repair genome will be so small that its progressive deterioration is unlikely to make much difference to the rate at which the vegetative Ratchet turns.

Although there will be no repair Ratchet as such in sexual populations, the fact that mutations in the repair system will be nearly neutral implies that mutations which erode the capacity for self-repair will often drift to fixation. It is natural to suspect that there may be mechanisms to prevent this. One possibility is that repair genes may be pleiotropic, contributing directly to vegetative function. They could then be selected on the basis of their vegetative consequences, and the repair load would be reduced because the selection coefficient attached to repair mutations would be much greater.

A second possibility is more speculative, and concerns the possible nature of endogenous systems of repair for the repair genome itself. Man-made error-correcting codes often utilize three (or more) copies of a message. All three copies are compared syllable by syllable. If a discrepancy is found, it is assumed that the message which differs from the other two is corrupt, and it is altered so as to bring all three into agreement. A three-stranded nucleic acid molecule could be repaired in the same way, and I presume that physical constraints have prevented the evolution of such an obvious system. It might still be possible to correct single-strand errors by using three copies of a gene on the same strand. Suppose that a gene were copied three times, the copies A, B and C occurring in sequence on the same strand. The simplest way of maintaining sequence homogeneity would be first to replicate the strand and then to shift the daughter strand relative to the parent strand, so that B on the original strand is now paired with A on the daughter strand, and C is

Figure 58. The time taken for the repair Ratchet to complete 100 turns. The contours represent this time for different values of k, given that each mutation in the repair genome increases the rate of mutation by a factor $(1+k)$, as a fraction of the time taken when $k = 0$.

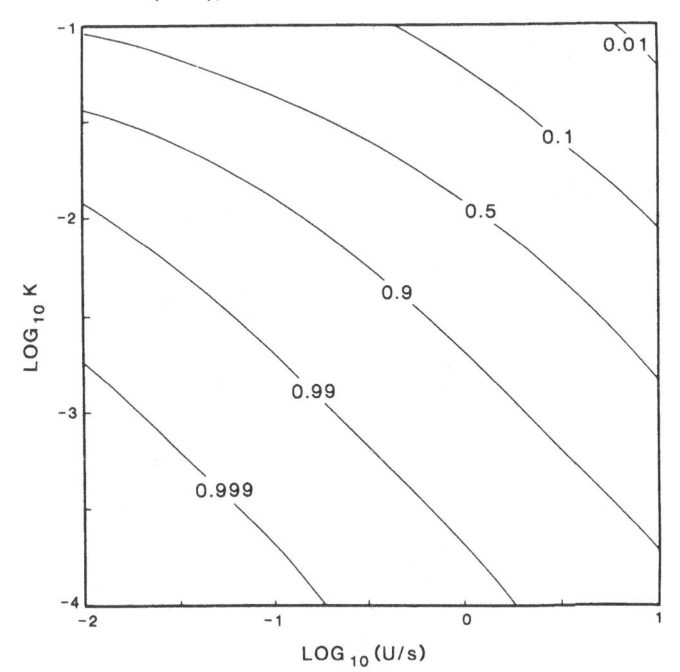

likewise paired with B (Figure 59). The two paired sequences B/A and C/B are now examined base by base. If a mismatch is found during the B/A comparison, a signal – say S_{BA} – is generated; a mismatch found during the C/B comparison generates a distinguishable signal S_{CB}. There are only four possible outcomes of the test, each leading to an unambiguous instruction for gene conversion. If S_{BA} but not S_{CB} is generated, A should be converted to B; if S_{CB} but not S_{BA} is generated, C should be converted to B; if both S_{CB} and S_{BA} are generated, then B should be converted to A or C; if neither S_{BA} nor S_{CB} are generated then no conversion is necessary. This scheme is entirely imaginary. I would not suggest it, except that the chain of thought I have followed above strongly suggests the desirability of some device by which the repair genome of sexual organisms can itself be repaired, independently of recombination and subsequent selection.

Figure 59. An hypothetical way of using three copies of a gene to preserve sequence homogeneity. Using 0 to indicate the absence of a signal, the appropriate responses to the results of the comparison would be:

S_{BA} and 0 : convert A to B

0 and S_{CB} : convert C to B

S_{BA} and S_{CB}: convert B to A or C

0 and 0 : no conversion

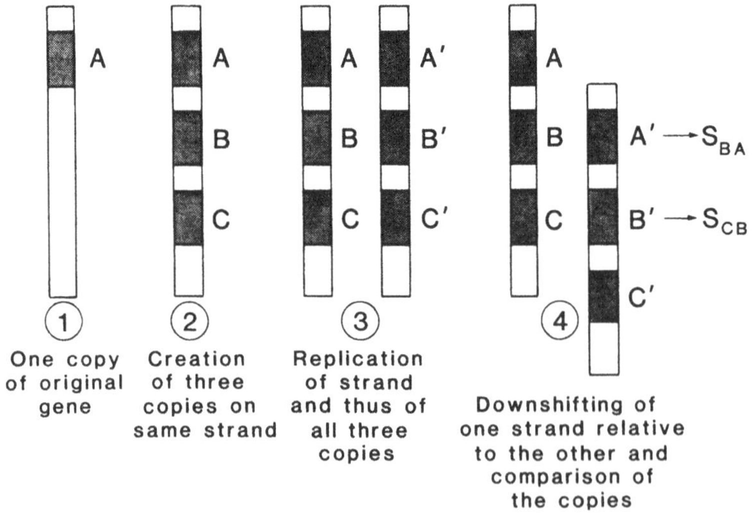

A
A B C
A B C A′ B′ C′
A B C A′ → S_{BA} B′ → S_{CB} C′

1 One copy of original gene
2 Creation of three copies on same strand
3 Replication of strand and thus of all three copies
4 Downshifting of one strand relative to the other and comparison of the copies

11

The function of sex

The operation of the Ratchet represents a serious threat to the integrity of germ lines which have no means of exogenous repair. In many circumstances this threat cannot be parried, and long-continued asexual propagation will eventually end in decline and extinction. The fate of isolate cultures is the *vera causa* of this process. This leads us to interpret sex as being maintained by most organisms – all those which do not have enormous populations or very small genomes – because of its role in containing the otherwise inexorable increase of mutational load.

The sort of evolutionary process envisaged by this argument is, roughly speaking, one of group selection. Competition between two reproductively isolated populations, one sexual and the other asexual, will be decided in favour of the sexual population, because the performance of its asexual competitor will gradually decay. It is probably necessary that they should be isolated: if asexual individuals occasionally produce reduced eggs which are fertilized by sexual males, then a gene which codes for almost completely obligate parthenogenesis will have an opportunity to become associated with lightly-loaded genomes. This is the force of Shield's (1982) arguments for the importance of partial inbreeding. I have described above how effective rather low rates of recombination will be in populations of moderate size. One example of such a process would be species selection: we are unlikely to observe asexuality, at least among relatively large organisms, because most obligately asexual species persist only for geologically brief spans of time. However, a similar process can operate even within an outcrossing sexual population, provided that genes coding for recombination are recessive to alleles which suppress recombination. Unloaded genomes can then be generated only by recessive homozygotes, and the recombination genes will spread as a result of the smaller mutational load with which they are associated. Felsenstein and

Yokoyama (1976) confirmed this inference by simulation, finding that high-recombination alleles tended to be fixed in populations exposed to deleterious mutation only if they were recessive. The limitation of this argument is that it holds only when the dominant gene completely suppresses recombination.

This interpretation enables us to make sense of a number of the leading features of parthenogenesis. The most important is perhaps its taxonomic distribution: the taxonomically isolated position of most groups of obligately parthenogenetic animals suggests that they are the short-lived offshoots of sexual stocks (Maynard Smith 1978, Bell 1982). Among protists, populations are often very large and the Ratchet will be less effective, so that the greater frequency of asexuality among smaller organisms is comprehensible. Further, the most prominent cytogenetic correlate of asexuality in metazoans is polyploidy: most parthenogenetic animals are polyploid, and most polyploid animals are parthenogenetic (Bell 1982). This might be because polyploidy slows down the initial stages of mutation-accumulation.

If these patterns do indeed reflect the general operation of the Ratchet, the process should be detectable, and could be studied in more detail, in contemporary systems which are for one reason or another very sensitive to the Ratchet. Isolate lines are only one example; it would be very instructive to conduct more such studies in organisms other than ciliates. Mitochondria, whose excessive variability I have interpreted above in terms of the Ratchet, are another case. More generally, the Ratchet will operate in any segment of the genome where recombination is suppressed, even in outcrossed sexual organisms. We would expect to find an accumulation of deleterious mutations on the Y chromosome, whose genetic inertness has been interpreted by Charlesworth (1978) as a device for silencing a Ratchetted genetic element, and within inversions, such as those actually used by Mukai to estimate the mutation rate of viability modifiers.

A very pretty demonstration of a Ratchet in the fish *Poeciliopsis* was published by Leslie and Vrijenhoek (1980). Some female *Poeciliopsis monacha* are hybridogenetic: they produce haploid eggs which are fertilized by males of a related species, and the daughters developing from these zygotes express both maternal and paternal factors. However, during oogenesis the entire paternal genome is excluded from the germ line, resulting in clonal transmission of the haploid maternal genome. We expect this genome to accumulate deleterious mutations (whose effect, in this particular case, may not be disastrous, because they are likely to be masked by genes from recombinant paternal genomes). By mating such a female with a diploid *P. monacha* male the hybridogenetic mechanism is to a great extent broken down, and progeny diploid for the *monacha* genome obtained. Whereas outbred sexual diploid *P. monacha* have mortality rates of less than 20%, the reconstituted diploids which include a clonally propagated maternal genome had an overall mortality of almost 30%. This strongly suggests the accumulation of a substantial number of incompletely recessive deleterious mutations in the clonal line of descent. A rather similar result has been obtained in the crustacean *Daphnia pulex* by David Innes and Paul Hebert (unpublished manuscript). Some populations of this species are heterogonic while others are obligately apomictic. The obligately apomictic females occasionally produce males, which are fully fertile and can be mated with females from the sexual stocks. The substantially elevated mortality of the F_2 generation of such crosses suggests the homozygous expression of deleterious mutations that have accumulated in the obligately asexual lines, especially since it is the paternal homozygote which generally appears to be deficient at electrophoretically characterized marker loci.

But the Ratchet cannot be the whole story. While the distribution of asexuality among metazoans is patchy, it is by no means haphazard. Sex is more frequent in marine than in freshwater habitats, at low rather than at high latitudes, and generally in old, stable, complex environments; sex generally replaces asexuality at high population density, and can usually be elicited in laboratory cultures of heterogonic organisms by crowding or starvation; propagules which disperse actively are almost always sexually produced. Such striking and widespread ecological patterns, which I have reviewed at length elsewhere (Bell 1982), do not seem to be predicted by the Ratchet.

Recombination creates variation with respect to the number of deleterious elements in the genome, so that selection can extract uncontaminated lines of descent which remain adapted in a uniform environment. But 'deleterious' is a laboratory concept, and it is only a laboratory worker who would imagine that environments are uniform. There are indeed grossly disfunctional mutations which will be severely deleterious in any environment in which the organism can survive at all. But the Ratchet is concerned with mutations of small effect. It will often be the case that, while deleterious in some circumstances, such mutations are mildly beneficial in others. Thus, if we measure the fitness of a series of genotypes created by mutation and recombination, the rank-order of their fitnesses will differ according to the environment in which it is measured. This amounts to an interaction between the performance of a genotype and the nature of the environment; some part of the variance of fitness will be attributable neither to differences between genotypes within an environment, nor to differences in average performance between environments, but rather to the fact that genotypes which perform relatively poorly in some environments may be better than average in others. Such interactions are crucial to understanding why sex is associated with stable, complex systems.

Suppose that the environment, instead of being physically uniform, is instead an exceedingly complex patchwork, with important factors such as light, water availability or nutrient levels varying substantially on a scale comparable with the size and activity of the organism. An asexual clone can efficiently colonize one type of patch, but as it increases in numbers its members will compete with increasing intensity because of their identical requirements, thus limiting the spread of the clone as a whole. The members of a sexual lineage, on the other hand, will compete less intensely because of the variety of conditions they can exploit, and may therefore come to outnumber and eventually to replace the clone. The greater productivity of a diverse brood in a highly heterogeneous environment represents an hypothesis of the maintenance of sexuality that I have called the Tangled Bank (Bell 1982).

There is remarkably little information on the scale of ecological variability in natural environments. A start has been made by Tilman (1982), whose results strongly support the patchwork-quilt model I have suggested above. However, discrete spatial patches are not essential to the theory. Even in a uniform environment, for example, a clone whose reproduction is limited by nitrogen availability will be supplanted by a mixed population, some of whose members are limited by nitrogen and others by phosphorus (see Clarke 1972).

There are several lines of evidence which support the notion that the fitness of a genotype varies according to circumstances. The most direct comes from experiments in which a number of genotypes are grown in a variety of habitats in a factorial manner, which are commonplace in the agronomic literature. The theory requires that the variance attributable to the interaction between genotype and environment is a substantial fraction of the total genetic variance, and I have briefly reviewed a number of cases which suggest that this is often true for characters closely related to fitness (Bell 1985b). Two more subtle and risky experiments have also been tried. The first is to move plants from their home site to another location, comparing their performance with related plants which, though otherwise receiving the same treatment, have been put back into their home plot. It is almost invariably found that being moved to a foreign plot reduces fitness. The second experiment assumes a homogeneous, or randomly assigned, environment, while the genotypic composition of the population is varied. If related plants have similar ecological requirements, then mixtures should have greater productivity than pure stands. Again, a brief review of a mainly agronomic literature shows that mixtures are indeed usually superior (Bell 1985b). A particularly elegant variation of this design is to surround a target plant either by related or by unrelated neighbours. Both Allard and Adams (1969) and Antonovics and Ellstrand (1984) found that the performance of the targets was greater, on average, when their neighbours were not closely related to them.

The extensive genotype-environment interaction to which these experiments point will tend to protect genetic variation by generating frequency-dependent selection. This is because a genotype will be able to exploit patches or resources which are currently under-utilized, so long as it is rare; as it becomes more common the members of this genotype compete among themselves and their relative fitness drops. Formal genetic models of such a process, first developed by Levene (1953) and Maynard Smith (1966), are a commonplace of the population genetics literature. If variation is maintained, it is tempting to speculate that the same circumstances will favour recombination and sexuality as means of diversifying progeny and thus exploiting the general superiority of mixtures. However, it is by no means clear that this will be the case, though some preliminary theoretical work has been published by Williams (1975), Bell (1982), Price and Waser (1982) and others. So long as sexual and asexual populations remain reproductively isolated, no great difficulty arises; the main problem is that a gene causing almost obligate parthenogenesis but permitting occasional outcrossing will not be limited to a single genetic background, so that at equilibrium the environment may be occupied by a diverse collection of clones rather than by a sexual population. The snag is similar to that which presents the Ratchet from operating, in general, as a process of selection between individuals. Whether or not genes which increase the rate of recombination can be selected within populations through their effect in increasing the variance within families is at present doubtful.

The situation is quite different if the resources being competed for are renewable. Imagine an annual plant with two varieties: one requires high levels of nitrogen but little phosphorus, while the other has the converse requirements. If these two varieties were planted in a garden where some plots received extra nitrogen and others extra phosphorus, their performance would show the sort of genotype–environment interaction that I have discussed above. In the field, the situation is complicated by the fact that a plant which uses large quantities of nitrogen might substantially deplete the availability of nitrogen at the site where it is growing. If nitrogen is replenished only slowly, its progeny are themselves liable to experience a shortage of nitrogen. The fitness of the nitrogen-requiring genotype will then vary inversely with its frequency in the previous generation, or perhaps in a sequence of previous generations. I have already explained how differential resource utilization will often lead to frequency-dependent selection; the example I have sketched above shows that when the relevant resources are renewable the operation of this frequency-dependence will be lagged in time. The effect of a time-lag will usually be to destabilize the genetic equilibria to which frequency-dependent selection would otherwise give rise, so that the genes controlling (in this case) the utilization of different mineral nutrients, instead of segregating at stable intermediate frequencies, would instead oscillate through time. It is unlikely that single genes control most characters of ecological importance; more plausibly, characters such as the response to nutrient availability will be influenced by genes at many loci. This immediately suggests that recombination will evolve because it offers the only rapid way of passing from one multilocus genotype to another. More formally, when the fitnesses of multilocus genotypes oscillate through time, a gene causing recombination will spread because it can become associated, through crossing-over, with the favoured genotype in each generation. Mathematical models which prove that high-recombination alleles may spread in populations exposed to time-lagged frequency-dependence have been published by Hutson and Law (1981) and Bell (1982). The process is pure individual selection; it is not necessary that genes which suppress and elevate recombination should define genetically isolated subpopulations.

A special case of this process is a system in which two organisms each constitute a renewable resource for one another, such as competitors, predator and prey, or pathogen and host. Provided that their interaction is sufficiently specific and intense, such mutual antagonists are locked into an endless arms-race with one another. To respond swiftly to the counteradaptation of an antagonist requires a means of breaking up old, discredited genotypes and building more effective ones from the pieces. The genetic instability of host–pathogen systems in particular, recognized by Haldane (1949) and Clarke (1979), may provide an opportunity for sexuality and recombination to evolve (Jaenike 1978, Hamilton 1980, Bremermann 1980, Tooby 1982, Bell 1982). Bell and Maynard Smith (1987) have recently analysed the coevolution of host and pathogen as a cyclical game, and show that under very general conditions the host-pathogen interaction will create powerful short-term individual selection for high rates of recombination.

There is as yet little direct evidence for what I have called the Red Queen interpretation of sexuality. It is roughly consistent with broad comparative trends, since species diversity and thus the incidence of antagonistic interactions between species is likely to be greater in the old, stable, complex environments where sex predominates. Somewhat more direct evidence is provided by the implication of pathogens in many cases where genotypic mixtures have been found to outyield pure strains (Bell 1985b; see especially Barrett 1981 and Wolfe *et al.* 1981). This is currently a rapidly developing field of enquiry in which very promising results have begun to emerge (e.g. Burt and Bell 1987, Lively 1987). There is, however, a great deal more work to be done before the Red Queen has a secure empirical foundation.

To sum up, a comprehensive interpretation of sexuality seems to require three distinct theories.

The first is that recombination functions as an endogenous repair system, providing a means of correcting otherwise lethal damage to the genetic material. This accounts for the origin, and of course the subsequent maintenance, of the molecular machinery required to shift genetic material from one strand of DNA to another.

Secondly, the recombination of genetic material between different lineages – sexual outcrossing – functions as an exogenous repair mechanism, without which large and complex organisms cannot persist through geological time. The old theory that sex rejuvenates the aging germ line is true: sex is part, probably much the most important part, of an exogenous repair system which functions by creating variance on which selection can act effectively to reduce mutational load. Organisms without sex are often doomed to rapid extinction, not only because an asexual germ line will accumulate deleterious mutations, but also because it will tend to accumulate them at an ever-increasing rate as a consequence of damage to endogenous repair systems. The Ratchet therefore identifies an essential condition for the existence of much of the living world. At the same time, it provides an explanation of several major features of sexuality, including the taxonomic distribution of parthenogenesis and its association with small size and polyploidy, besides making predictions about the behaviour of particular systems such as isolate lines, mitochondria and inversions.

What the Ratchet leaves unexplained are the ecological correlates of contemporary sexual systems, and in particular the association of sexuality in metazoans and metaphytes with old, stable and complex environments. These patterns lead to a third hypothesis, which points out that that recombination of deleterious genes has the necessary side-effect of creating new combinations of genes which although deleterious in some circumstances are beneficial in others. Sexual diversification can then be favoured because it increases the efficiency with which offspring can exploit a heterogeneous environment. The advantage of sex is clearest when individuals are competing for renewable resources, when it will often be the case that recombination provides the only way of passing rapidly between the different multilocus genotypes favoured in different generations. These ideas enable us to predict the magnitude of genotype–environment interactions, and the results of experiments such as reciprocal transplantation and the comparison of mixtures with pure lines. They also lead us to suspect that antagonistic interactions between species, and especially between pathogens and their hosts, may play an important part in the evolution of sex.

Taken together, these three ideas serve both to organize our knowledge of the comparative biology of sexual systems and to predict the results of experiments in the field and in the laboratory. Each seems to be necessary; all three together may be sufficient, since I do not know of any major phenomena which they leave unexplained. In this way, I think, the old wrangle between Ehrenberg, Weismann, Maupas, Calkins, Woodruff, Jennings and my other heroes can at last be resolved.

BIBLIOGRAPHY

Absher, P.M. and Absher, R.G. (1976). Clonal variation and aging of diploid fibroblasts. *Exp. Cell Res.*, **103**, 247.

Ackert, J.E. (1916). On the effect of selection in *Paramecium. Genetics*, **1**, 387–405.

Allard, R.W. and Adams, J. (1969). Population studies in predominantly self-fertilizing species. XIII. Intergenotypic competition and population structure in barley and wheat. *Amer. Nat.* **103**, 621–645.

Allen, S. and Gibson, J. (1972). Genetics of *Tetrahymena*. In *Biology of Tetrahymena*, ed. A.M. Elliott, pp. 307–73. Stroudsberg, Pennsylvania: Bowden Hutchinson and Ross.

Ammermann, D. (1971). Morphology and development of the macronuclei of the ciliates *Stylonichia mytilus* and *Euplotes aediculatus. Chromosoma*, **33**, 209–38.

Ammermann, D., Steinbruck, D., von Berger, L. and Hennig, W. (1974). The development of the macronucleus in the ciliated protozoan, *Stylonichia mytilus. Chromosoma*, **45**, 401–29.

Antonovics, J. and Ellstrand, N. (1984). Experimental studies of the evolutionary significance of sex. I. A test of the frequency-dependent selection hypothesis. *Evolution*, **38**, 103–15.

Ashby, E., Wangermann, E. and Winter, E.J. (1984). Studies in the morphogenesis of leaves. III. Preliminary observations on vegetative growth in *Lemna minor. New Phytol.*, **47**, 374–81.

Austin, M.L. (1927). Studies on *Uroleptus mobilis*. I. An attempt to prolong the life cycle. *J. exp. Zool.*, **49**, 149–216.

Baitsell, G.A. (1912). Experiments on the reproduction of the hypotrichous infusoria. I. Conjugation between closely-related individuals of *Stylonichia pustulata. J. exp. Zool.*, **13**, 47–77.

Baitsell, G.A. (1914). *Ibid.* II. A study of the so-called life cycle in *Oxytricha fallax* and *Pleurotricha lanceolata. J. exp. Zool.*, **16**, 211–35.

Balbiani, E.G. (1860). Observations et experiences sur les phenomenes de reproduction fissipare chez les infusoires ciliés. *C. R. Acad. Sci. Paris*, **50**, 1191.

Balbiani, E.G. (1882). Les Protozoaires. *J. Microgr.*, vols 5 and 6.

Banta, A.M. (1914). One hundred parthenogenetic generations of *Daphnia* without sexual forms. *Proc. Soc. exp. Biol. Med.*, **11**, 180–2.
Banta, A.M. (1915). The effects of long-continued parthenogenetic reproduction (127 generations) upon daphnids. *Science*, **41**, 442.
Banta, A.M. and Wood, T.R. (1937). The accumulation of recessive physiological mutations during long-continued parthenogenesis. *Genetics*, **22**, 183–4.
Barrett, J.A. (1981). The evolutionary consequences of monoculture. In *The Genetic Consequences of Man-Made Change*, ed. J.A. Bishop and L.M. Cook, pp. 209–48. London: Academic Press.
Beers, C.D. (1926). The life cycle in the ciliate *Didinium nasutum*. *J. Morphol. physiol.*, **42**, 1–21.
Beers, C.D. (1929). On the possibility of indefinite reproduction in the ciliate *Didinium* without conjugation or endomixis. *Amer. Natur.*, **63**, 125–9.
Beers, C.D. (1933). Diet in relation to depression and recovery in the ciliate *Didinium nasutum*. *Arch. Protistenk*, **79**, 101–18.
Belar, K. (1922). Untersuchungen an *Actinophrys sol* Ehrenberg. I. Die Morphologie des Formwechsels. *Arch. Protistenk*, 46, 1–96.
Belar, K. (1924). Untersuchungen an *Actinophrys sol* Ehrenberg. II. Beitrage sur Physiologie des Formwechsels. *Arch. f. Protistenk.*, **48**, 371–434.
Beguet, B. (1972). The persistence of processes regulating the level of reproduction in the hermaphroditic nematode *Caenorhabditis elegans*, despite the influence of parental ageing, over several consecutive generations. *Exp. Gerontol.*, **7**, 209–18.
Bell, G. (1982). *The Masterpiece of Nature*. London: Croom-Helm, Berkeley: University of California Press.
Bell, G. (1984). Measuring the cost of reproduction. II. The correlation structure of the life tables of five freshwater invertebrates. *Evolution*, **38**, 314–26.
Bell, G. (1985a). Evolutionary and non-evolutionary theories of senescence. *Amer. Natur*, **124**, 600–3.
Bell, G. (1985b). Two theories of sex and variation. *Experientia*, **10**, 1235–45.
Bell, G. (1985c). The origin and early evolution of germ cells, as illustrated by the Volvocales. In *The Origin and Evolution of Sex*, ed. H. Halvorson and A. Monroy, pp. 221–56. New York: Liss.
Bell, G. and Koufopanou, V. (1986). The cost of reproduction. *Oxford Surveys in Evolutionary Biology*, **3**, 83–131.
Bell, G. and Maynard Smith, J. (1987). Short-term selection for recombination among mutually antagonistic species. *Nature*, **327**, 66–8.
Bennett, M.D. (1971). The duration of meiosis. *Proc. R. Soc.*, **B178**, 277–99.
Bernstein, H., Byerly, H.C., Hopf, F.A. and Michod, R.E. (1985a). The evolutionary role of recombinational repair and sex. *Int. Rev. Cytology*, **96**, 1–28.
Bernstein, H., Byerly, H.C., Hopf, F.A. and Michod, R.E. (1985b). Genetic damage, mutation and the evolution of sex. *Science*, **229**, 1277–81.
Bernstein, H., Byerly, H.C., Hopf, F.A. and Michod, R.E. (1985c). DNA repair and complementation: the major factors in the origin and maintenance of sex. In *The Origin and Evolution of Sex*, ed. H. Halvesam and A. Monroy, pp. 29–45. New York: Liss.

Blueweiss, L., Fox, H., Kudzma, V., Nakashima, D., Peters, R. and Sams, S. (1978). Relationship between body size and some life history parameters. *Oecologia*, **37**, 257–72.

Bremermann, H.J. (1980). Sex and polymorphism as strategies in host-pathogen interactions. *J. theoret. Biol.*, **87**, 671–702.

Brien, P. (1953). La perennité somatique. *Biol. Rev.*, **28**, 308–49.

Burnet, A. (ed) (1973). *The Biology of Hydra*. New York: Academic Press.

Burt, A. and Bell, G. (1987). Mammalian chiasma frequencies as a test of two theories of recombination. *Nature*, **326**, 803–5.

Butschli, O. (1873). Einiges uber Infusorien. *Arch. mikr. Anat.*, **9**, 657–8.

Butschli, O. (1876). Studien uber die Entwicklungsvorgange der Eizelle, die Zellteilung ind die Conjugation der Infusorien. *Abh. Senchkenberg Naturf. Ges.*, **10**, 213–452.

Caldwell, L. (1933). The production of inherited diversities at endomixis in *Paramecium aurelia*. *J. exp. Zool.*, **66**, 371–407.

Calkins, G.N. (1902). Studies on the life history of Protozoa. I. The life cycle of *Paramecium caudatum*. *Arch. Entwicklungs-Mech. Org.*, **15**, 139–86.

Calkins, G.N. (1903). Studies on the life history of Protozoa. III. The six hundreth and twentieth generation of *Paramecium caudatum*. *Biol. Bull.*, **3**, 192–205.

Calkins, G.N. (1904). Studies on the life history of Protozoa. IV. Death of the A-Series. Conclusions. *J. exp. Zool.*, **12**, 423–61.

Calkins, G.N. (1912). The paedogamous conjugation of *Blepharisma undulans* St. *J. Morphol.*, **23**, 667–87.

Calkins, G.N. (1915). Cycles and rhythms and the problems of 'immortality' in *Paramecium*. *Amer. Natur.*, **49**, 65–76.

Calkins, G.N. (1919). *Uroleptus mobilis* Engelm. II. Renewal of vitality through conjugation. *J. exp. Zool.*, **29**, 121–56.

Calkins, G.N. (1920). *Uroleptus mobilis* Engelm. III. A study in vitality. *J. exp. Zool.*, **31**, 287–305.

Calkins, G.N. (1925). *Uroleptus mobilis*. V. The history of a double organism. *J. exp. Zool.*, **41**, 191–213.

Calkins, G.N. (1926). *Biology of the Protozoa*. Philadelphia.

Calkins, G.N. and Gregory, L.H. (1913). Variations in the progeny of a single exconjugant of *Paramecium caudatum*. *J. exp. Zool.*, **15**, 467–525.

Calkins, G.N. and Lieb, C.C. (1902). Studies on the life history of Protozoa. II. The effect of stimuli on the life cycle of *Paramecium caudatum*. *Arch. Protistenk*, **1**, 355–71.

Cameron, I.L. (1972). Minimum number of cell doublings in an epithelial population during the life span of the mouse. *J. Gerontol.*, **27**, 157.

Carrell, A. (1912). On the permanent life of tissues outside of the organism. *J. exp. Med.*, **15**, 516.

Charlesworth, B. (1978). Model for evolution of Y chromosomes and dosage compensation. *Proc. Nat. Acad. Sci. USA*, **75**, 5618–22.

Child, C.M. (1911). A study of senescence and rejuvenescence based on experiments with *Planaria dorotocephala*. *Arch. Entweicklungs-Mech. Org.*, **31**, 537–616.

Child, C.M. (1914). Asexual breeding and prevention of senescence in *Planaria velata*. *Biol. Bull.*, **26**, 286–93.

Child, C.M. (1915). *Senescence and Rejuvenescence*. Chicago.
Child, C.M. (1916). Age cycles and other periodicities in organisms. *Proc. Amer. Philos. Soc.*, **55**, 330–9.
Clarke, B.C. (1972). Density-dependent selection. *Amer. Nat.*, **106**, 1–13.
Clarke, B.C. (1979). The evolution of genetic diversity. *Proc. R. Soc.* **B205**, 453–74.
Cohen, B.M. (1933). The effect of conjugation within a clone of *Euplotes patella*. *Genetics*, **19**, 25–39.
Comfort, A. (1953). *The Biology of Senescence*. Edinburgh: Churchill Livingstone.
Danes, B.S. (1971). Progeria: a cell culture study on aging. *J. clin. Invest.*, **50**, 2000.
Danielli, J.F. and Muggleton, A. (1959). Some alternative states of amoeba, with special reference to life span. *Gerontologia*, **3**, 76.
Dawson, J.A. (1919). An experimental study of an amicronucleate *Oxytricha*. *J. exp. Zool.*, **29**, 473–513.
Dawson, J.A. (1926). The 'life cycle' of *Histrio complanatus*. *J. exp. Zool.*, **46**, 345–53.
Dell'Orco, R.T., Mertens, J.G. and Kruse, P.F. (1974). Doubling potential, calendar time and donor age of human diploid cells in culture. *Exp. Cell. Res.*, **84**, 363.
DeNoyelles, F. (1971). PhD thesis, Cornell University.
Diller, W.F. (1936). Nuclear reorganization processes in *Paramecium aurelia*, with descriptions of autogamy and 'hemixis'. *J. Morphol.*, **59**, 11–66.
Doolittle, W.F. and Sapienza, C. (1980). Selfish genes, the phenotype paradigm and genome evolution. *Nature*, **284**, 601–3.
Dujardin, F. (1841). *Histoire Naturelle des Zöophytes Infusoires, Comprenant la Physiologie et la Classification de ces Animaux*. Paris.
Dupy-Blanc, J. (1969). Etude cytophotométrique des teneaurs en ADN des micronucléus de *Paramecium caudatum* au cours de la conjugaison et pendant les differentiations des 'Anlagen' en macronucléus. *Protistologica*, **5**, 239–48.
Earle, W.R. (1943). Production of malignancy in vitro. IV. The mouse fibroblast cultures and changes seen in the living cells. *J. Nat. Cancer Inst.*, **4**, 165.
Ebeling, A.H. (1913). The permanent life of connective tissue outside of the organism. *J. exp. Med.*, **17**, 273.
Edney, R.B. and Gill, R.W. (1968). Evolution of senescence and specific longevity. *Nature*, **220**, 281–2.
Ehrenberg, C.G. (1838). *Die Infusionstier als Vollkommene Organismen*. Leipzig.
Engelmann, T.W. (1862). Zur Naturgeshichte der Infusionsthiere. *Z. Wiss. Zool.*, **11**, 1–47.
Engelmann, T.W. (1876). Uber Entwicklung und Fortpflanzung der Infusorien. *Morph. Jahrbuch*, **1**, 573–634.
Enriques, P. (1903). Sulla cosi detta 'degenerazione senile' dei protozoa. *Monit. Zool. Ital.*, **14**, 351–7.
Enriques, P. (1916). Duemila cinquecento generazione in un infusorio, senze coniugazione ne partenogenesi, ne depressioni. *Rend. Sess. R. Accad. Sci. Inst. Bologna Ser. 7*, **T4**, 1–12.

Erdmann, R. (1920). Endomixis and size variations in pure lines of *Paramecium aurelia*. *Proc. Soc. exp. Biol. Med.*, **16**, 60–5.
Erdmann, R. and Woodruff, L.L. (1916). The periodic reorganization process in *Paramecium caudatum*. *J. exp. Zool.*, **20**, 59–97.
Evans, G.M., Rees, H., Snell, C.L. and Sun, S. (1972). The relationship between nuclear DNA amount and the duration of the mitotic cycle. *Chromosomes Today*, **3**, 24–31.
Fabre-Domergue, P. (1889). *Annales Microgr.*, **2**, 237. (Not seen.)
Fauré-Fremiet, E. (1953). L'hypothèse de la sénescence et les cycles de réorganisation nucléaire chez les Ciliés. *Rev. Suisse Zool.*, **60**, 426–38.
Felsenstein, J. (1974). The evolutionary advantage of recombination. *Genetics*, **78**, 737–56.
Felsenstein, J. and Yokoyama, S. (1976). The evolutionary advantage of recombination. I. Individual selection for recombination. *Genetics*, **83**, 845–59.
Frankel, J. (1973). Dimensions of control of cortical patterns in *Euplotes*: the role of pre-existing structure, the clonal life-cycle, and the genotype. *J. exp. Zool.*, **183**, 71–94.
Fukushima, S. (1974). Effect of X-irradiations on the clonal lifespan and fission rate in *Paramecium aurelia*. *Exp. Cell. Res.*, **84**, 267–70.
Galadjieff, M. and Metalnikov, S. (1935). L'immortalité de la cellule. Vingt-deux ans de culture d'Infusoires sans conjugaison. *Arch. Zool. exp. gén.*, **75**, 331–52.
de Garis, C.F. (1927). A genetic study of *Paramecium caudatum* in pure lines through an interval of experimentally induced monster formation. *J. exp. Zool.*, **49**, 133–47.
de Garis, C.F. (1928). The effects of anterior and posterior selections on fission rate in pure lines of *Paramecium caudatum*. *J. exp. Zool.*, **50**, 1–14.
Gelber, J. (1938). The effect of shorter than normal interendomictic intervals on mortality after endomixis in *Paramecium aurelia*. *Biol Bull.*, **74**, 244–6.
Gerassimoff, J.J. (1901). Ueber die Kerntheilung bie *Actinosphaerium eichhorni*. *Jenais Zeitschr*, **17**, 490–518.
Gerassimoff, J.J. (1902). Die Abhangigkeit der Grosse der Zelle von der Menge ihre Kernmasse. *Z. allg. Physiol.*, **1**, 220–58.
Gey, G.O. and Gey, M.K. (1936). The maintenance of human normal cells and tumour cells in continuous culture. I. Preliminary report: cultivation of mesoblastic tumours and normal tissue and notes on methods of cultivation. *Amer. J. Cancer*, **27**, 45.
Gibson, I. and Martin, N. (1971). DNA amounts in the nuclei of *Paramecium aurelia* and *Tetrahymena pyriformis*. *Chromosoma*, **35**, 374–82.
Goetsch, W. (1922). Beitrage zum Unsterblichkeits problem der Metazoen. II. Lebensdauer und geschlechtilige Fortpflanzung bei *Hydra*. *Biol. Zentralblatt*, **42**, 231–40.
Goetsch, W. (1925). Versuche und Beobachtungen uber die Biologie der Hydriden. *Biol. Zentralblatt*, **45**, 321–52, 385–417.
Goldstein, S. (1969). Lifespan of cultured cells in progeria. *Lancet*, 1969, 424.
Goldstein, S. and Singal, D.P. (1974). Senescence of cultured human fibroblasts: mitotic versus metabolic time. *Exp. Cell. Res.*, **88**, 359.

Gregory, L.H. (1909). Observations on the life history of *Tillina magna*. *J. exp. Zool.*, **6**, 383–431.

Hackett, H.E., Boss, M.L. and Dijkman, M.J. (1963). Regeneration and proportional life span extension of *Acetabularia crenata* Lamouroux. *J. Gerontol.*, **18**, 331–4.

Hadorn, E. (1967). Dynamics of determination. In *Major Problems in Developmental Biology*, ed. M. Locke, 25th Symp. Soc. Dev. Biol. New York: Academic Press.

Haigh, J. (1978). The accumulation of deleterious genes in a population – Muller's Ratchet. *Theoret. Pop. Biol.*, **14**, 251–67.

Haldane, J.B.S. (1949). Disease and evolution. *La Ricerca Scientifica* (*Suppl.*) **19**, 68–76.

Hamilton, W.D. (1980). Sex versus non-sex versus parasite. *Oikos*, **35**, 282–90.

Hammerling, J. (1924). Die ungeschlechtiliche Fortpflanzung und Regeneration bei *Aelosoma hemprichii. Zool. Jahrbuch*, **41**, 581–655.

Hartmann, M. (1917). Untersuchungen uber die Morphologie und Physiologie des Formwechsels (Entwicklung, Fortpflanzung und Vererbung) der Phytomonadinen (Volvocales). II. Uber die dauernds rein agame Zuchtung von *Eudorina elegans* und ihre Bedeutung fur das Befruchtungs- und Todproblem. *Sitz Berlin preuss Akad. Wiss. Phys.-Math. Kl*, 1917. 760–76.

Hartmann, M. (1921). *Ibid.* III. Die dauernde agame Zucht von *Eudorina elegans*, experimentelle Beitrage zum Befruchtungs-und todproblem. *Arch. Protistenk*, **43**, 223–86.

Hartmann, M. (1922). Uber den dauernden Ersatz der ungeschlechtlichen Fortpflanzung durch fortgesetzte Regeneration. *Biol. Zentralbl.*, **42**, S. 364.

Hartmann, M. (1924). Der Ersatz der Fortpflanzung von Amoeben durch Fortgesetzte Regenerationen. *Arch. Protistenk*, **49**, 447–64.

Hase, A. (1909). Uber die duetschen Susswasser-Polypen *Hydra fusca* L. *Hydra grisea* L. und *Hydra viridis* L. *Arch. Rassen- u Gesellschafts-Biologie*, **6**, 721–53.

Hayflick, L. (1965). The limited in vitro lifetime of human diploid cell strains. *Exp. Cell. Res.*, **37**, 614–36.

Hayflick, L. and Moorhead, P.S. (1961). The serial cultivation of human diploid cell strains. *Exp. Cell. Res.*, **25**, 585.

Hegner, R.W. (1919). Heredity, variation and the appearance of diversities during the vegetative reproduction of *Arcella dentata. Genetics*, **4**, 95–150.

Heller, R. and Maynard Smith, J. (1979). Does Muller's Ratchet work with selfing? *Genet. Res. Cambridge*, **32**, 289–93.

Hertwig, R. (1884). Ueber die Kerntheilung bie *Actinosphaerium eichhorni. Jenais Zitschrift*, **17**, 490–518.

Hertwig, R. (1889). Ueber die conjugationen der Infusorien. *Abhandl. konigl. bayr. Akad. der Wiss.*, **17**, 151–233.

Hertwig, R. (1903). Ueber Korrelation von Zell- und Kerngrosse und ihre Beduetung fur die geschlechtliche Differenzierung und die Teilung der Zelle. *Biol. Zentralblatt*, **23**, 49–62, 108–19.

Hutson, V. and Law, R. (1981). Evolution of recombination in populations experiencing frequency-dependent selection with time delay. *Proc. R. Soc.*, **B213**, 345–59.

Iwamura, Y., Sakai, M., Mita, T. and Muramatsu, M. (1979). Unequal gene amplification and transcription in the macronucleus of *Tetrahymena pyriformis*. *Biochemistry*, **18**, 5289–94.

Jaenike, J. (1978). An hypothesis to account for the maintenance of sex within populations. *Evol. Theory*, **3**, 191–4.

Januszko, M. (1971). Growth of green algae in fertilized and unfertilized fish fry ponds. *Polskie Arch. Hydrobiol.*, **18**, 129–49.

Jennings, H.S. (1908). Heredity, variation and evolution in Protozoa. II. Heredity and variation in size and form in *Paramecium*, with studies of growth, environmental action and selection. *Proc. Amer. Philos. Soc.*, **47**, 393–546.

Jennings, H.S. (1913). The effect of conjugation in *Paramecium*. *J. exp. Zool.*, **14**, 279–391.

Jennings, H.S. (1916). Heredity, variation and the results of selection in the uniparental reproduction of *Difflugia corona*. *Genetics*, **1**, 407–534.

Jennings, H.S. (1929). Genetics of the Protozoa. *Bibl. Genetica*, **5**, 105–330.

Jennings, H.S. (1944a). *Paramecium bursaria*: life history. I. Immaturity, maturity and age. *Biol. Bull.*, **86**, 131–45.

Jennings, H.S. (1944b). *Ibid.* II. Age and death of clones in relation to the results of conjugation. *J. exp. Zool.*, **96**, 17–52.

Jennings, H.S. (1944c). *Ibid.* III. Repeated conjugations in the same stock at different ages, with and without inbreeding, in relation to mortality at conjugation. *J. exp. Zool.*, **96**, 242–73.

Jennings, H.S. (1944d). *Ibid.* IV. Relation of inbreeding to mortality of exconjugant clones. *J. exp. Zool.*, **97**, 165–97.

Jennings, H.S. (1945). *Ibid.* V. Some relations of external conditions, past and present, to ageing and to mortality of exconjugants, with summary of conclusions on age and death. *J. exp. Zool.*, **98**, 15–31.

Jennings, H.S. and Lynch R.S. (1928). Age, mortality, fertility and individual diversities in the rotifer *Proales sordida* Gosse. I. Effect of the age of the parent on characteristics of the offspring. *J. exp. Zool.*, **50**, 345–407.

Jennings, H.S. Raffel, D., Lynch R.S. and Sonneborn, T.M. (1932). The diverse biotypes produced by conjugation within a clone of *Paramecium aurelia*. *J. exp. Zool.*, **62**, 363–408.

Johannsen, W. (1903). *Ueber Erblichkeit in Populationen ind in reinen Linien*. Jena.

Jollos, V. (1921). Experimentelle Protistenstudien. I. Untersuchungen uber Variabilitat und Vererbung bie Infusorien. *Arch. Protistenk*, **43**, 1–222.

Jones, A.R. (1974). *The Ciliates*. London: Hutchinson.

Joukowsky, D. (1898). Beitrage zur Frage nach den Bedingungen der Vermehrung und des Eintritts der Conjugation bei den Ciliaten. *Verhandl. Nat.-Med. Ver Heidelberg*, **6**, 17–42.

Kimball, R.F. and Gaither, N. (1954). *Genetics*, **39**, 977.

Kimura, A. (1957). Some problems of stochastic processes in genetics. *Ann. Math. Stat.*, **28**, 882–901.

King, C.E. (1967). Food, age and the dynamics of a laboratory population of rotifers. *Ecology*, **48**, 111–28.

Kirkwood, T.B.L. and Cremer, T. (1982). Cytogerontology since 1891: a reappraisal of August Weismann and a review of modern progress. *Human Genetics*, **60**, 101–21.

Kochert, G. (1973). Colony differentiation in green algae. In *Developmental Regulation*, ed. S.J. Coward, pp. 155–67. New York: Academic Press.
Kohn, R.R. (1981). Evidence against cellular aging theories. In *Testing the Theories of Aging*, ed. R.C. Adelman and G.S. Roth, pp. 221–31. Florida: CRC Press.
Kondrashov, A.S. (1984). Deleterious mutation as an evolutionary factor. I. The advantage of recombination. *Genet. Res. Cambridge*, **44**, 199–217.
Krooth, R.S., Shaw, M.W. and Campbell, B.K. (1964). A persistent strain of diploid fibroblasts. *J. Nat. Cancer Inst.*, **32**, 1031.
Lackey, J.B. (1929). Studies in the life histories of Euglenida. II. The life cycles of *Entosiphon sulcatum* and *Peranema trichophorum*. *Arch. Protistenk*, **67**, 128–56.
Lange, C.S. (1981). A possible explanation in cellular terms of the physiological ageing of the planarian. In *Biology of the Turbellaria: Experimental Advances II*. pp. 99–109, New York: MSS Information Corp.
Lansing, A.I. (1947). A transmissible, cumulative and reversible factor in aging. *J. Gerontol.*, **2**, 228–39.
Lansing, A.I. (1954). A nongenic factor in the longevity of rotifers. *Ann. New York Acad. Sci.*, **57**, 455–64.
Leslie, J.F. and Vrïjenhoek, R.C. (1980). Consideration of Muller's Ratchet mechanism through studies of genetic linkage and genomic compatibilities in clonally reproducing *Poeciliopsis*. *Evolution*, **34**, 1105–15.
Levene, H. (1953). Genetic equilibrium when more than one ecological niche is available. *Amer. Nat.*, **87**, 131–3.
Leith, A.G. (1984). The death of an amoeba. *Protozool.*, **31**, 177–180.
Lints, F.A. and Hoste, C. (1974). The Lansing effect revisited. I. Life-span. *Exp. Gerontol.*, **9**, 51–69.
Lively, C. (1987). Evidence from a New Zealand snail for the maintenance of sex by parasitism. *Nature*, **328**, 519–21.
Lynch, M. (1985). Spontaneous mutations for life history characters in an obligate parthenogen. *Evolution*, **39**, 804–18.
Lynch, R.S. and Smith, H.B. (1931). A study of the effects of modification of the culture medium upon length of life and fecundity in a rotifer, *Proales sordida*, with special relation to their heritability. *Biol. Bull.*, **60**, 30–59.
McHale, J.S., Mouton, M.L. and McHale, J.T. (1971). Limited culture lifespan of human diploid cells as a function of metabolic time instead of division potential. *Exp. Gerontol.*, **6**, 89.
McTavish, C. and Sommerville, J. (1980). Macronuclear DNA organization and transcription in *Paramecium primaurelia*. *Chromosoma*, **78**, 147–64.
Manning, J.T. (1983). Diploidy and Muller's ratchet. *Acta biotheoretica*, **32**, 289–92.
Manning, J.T. and Dickson, D.P.E. (1986). Asexual reproduction, polyploidy and optimal mutation rates. *J. theoret. Biol.*, **118**, 485–9.
Martin, G.M., Ogburn, C.E. and Sprague, C.A. (1981). Effects of age on cell division capacity. In *Aging: A Challenge to Science and Society*, ed. D. Dannon, N.M. Shock and M. Marcois, p. 124. Oxford Univ. Press.
Martin, G.M., Sprague, C.A. and Epstein. C.J. (1970). Replicative life span of cultivated human cells: effects of donor's age, tissue and genotype. *Lab. Invest.*, **23**, 86.

Mast, S.O. (1917). Conjugation and encystment in *Didinium nasutum* with especial reference to their significance. *J. exp. Zool.*, **23**, 335–59.
Mather, K. and Jinks, J.L. (1958). Cytoplasm in sexual reproduction. *Nature*, **182**, 1188–90.
Maupas, E. (1888). Recherches experimentales sur la multiplication des infusoires ciliés. *Arch. Zool. exp. gen.*, **6**, 165–277.
Maupas, E. (1889). La rejeunissement karyogamique chez les ciliés. *Arch. Zool. exp. gen.*, **7**, 149–517.
Maupas, E. (1919). Experiences sur la reproduction asexuelle des oligochetes. *Bull. Biol. France Belgique*, **53**, 150–60.
Maynard Smith, J. (1959). A theory of aging. *Nature*, **184**, 956–8.
Maynard Smith, J. (1966). Sympatric speciation. *Amer. Nat.*, **100**, 637–50.
Maynard Smith, J. (1978). *The Evolution of Sex*. Cambridge Univ. Press.
Meadow, N.D. and Barrows, C.H. (1971). Studies on aging in a bdelloid rotifer. II. The effects of various environmental conditions and maternal age on longevity and fecundity. *J. Gerontol.*, **26**, 302–9.
Medawar, P.B. (1952). *An Unresolved Problem in Biology*. London: H.K. Lewis.
Menhaus, A.J., de Jong, B. and ten Kate, L.P. (1971). Fibroblast culture in Werner's syndrome. *Human Genetics*, **13**, 244.
Melander, Y. (1963). Cited by Lange, op cit.
Metalnikov, S. (1919). L'immortalité des organismes unicellulaires. *Ann. Inst. Pasteur*, **33**, 817–35.
Metalnikov, (1922). Dix ans de culture des infusoires sans conjugaison. *C. R. Acad. Sci. Paris*, **175**, 776–8.
Metalnikov, S. (1924). *Immortalité et Rejeunissement dans la Biologie Moderne*. Paris.
Middleton, A.R. (1915). Heritable variations and the results of selection in the fission rate of *Stylonichia pustulata*. *J. exp. Zool.*, **19**, 4351–503.
Moody, J.L. (1912). Observations on the life history of two rare ciliates, *Spathidium spathula* and *Actinobulus radians*. *J. exp. Zool.*, **23**, 349–400.
Moore, G.E. and Sandberg, A.A. (1964). Studies on a human tumor cell line with a diploid karyotype. *Cancer*, **17**, 170.
Mukai, T. (1964). The genetic structure of natural populations of *Drosophila melanogaster*. I. Spontaneous mutation rate of polygenes controlling viability. *Genetics*, **50**, 1–19.
Mukai, T., Chigusa, S.T., Mettler, L.E. and Crow, J.F. (1972). Mutation rate and dominance of genes affecting viability in *Drosophila melanogaster*. *Genetics*, **72**, 335–55.
Muggleton, A. and Danielli, J.F. (1958). Ageing of *Amoeba proteus* and *Amoeba discoides* cells. *Nature*, **181**, 1738.
Muggleton, A. and Danielli, J.F. (1968). Inheritance of the 'life-spanning' phenomenon in *Amoeba proteus*. *Exp. Cell Res.* **49**, 116–20.
Muller, H.J. (1964). The relation of recombination to mutational advance. *Mutation Res.*, **1**, 2–9.
Nanney, D.L. (1959). Vegetative mutants and clonal senility in *Tetrahymena*. *J. Protozool.*, **6**, 171–77.
Nanney, D.L. (1980). *Experimental Ciliatology*. New York: Wiley.
von Neumann, J. (1966). *Theory of self-reproducing Automata*. Urbana: Univ. Illinois Press.

Orgel, L.E. (1963). The maintenance of the accuracy of protein synthesis and its relevance to aging. *Proc. Nat. Acad. Sci. US*, **49**, 517.
Orgel, L.E. (1973). Aging of clones of mammalian cells. *Nature*, **243**, 441.
Orgel, L.E. and Crick, F.H.C. (1980). Selfish DNA: the ultimate parasite. *Nature*, **284**, 604–7.
Parker, R.C. (1927). The effect of selection in a pedigree line of infusoria. *J. exp. Zool.*, **49**, 401–39.
Pearl, R. and Miner, J.R. (1935). Experimental studies on the duration of life. XIV. The comparative mortality of certain lower organisms. *Quart. Rev. Biol.*, **10**, 60–79.
Peters, R.H. (1983). *The Ecological Importance of Body Size*. Cambridge Univ. Press.
Petersen, W.A. (1927). The axial gradient in *Paramecium*. *Science*, **66**, 157–8.
Petursson, G., Coughlin, J.G. and Meylan, C. (1964). Long-term cultivation of diploid rat cells. *Exp. Cell. Res.*, **33**, 60.
Pierson, B.F. (1938). The relation of mortality after endomixis to the prior interendomictic interval in *Paramecium aurelia*. *Biol. Bull.*, **74**, 235–43.
Popoff, N. (1907). Depression der Protozoenzelle und der Geschlechtszellen der Metazoen. *Arch. f. Protistenk. Suppl.* **Bd 1**, 42–83.
Popoff, N. (1908). Experimentelle Zellstudien. I. *Arch Zellforsch*, **1**, 245–379.
Popoff, N. (1909). Experimentelle Zellstudien. II. Ueber die Zellgrosse, ihre Fixierung und Vererbung. *Arch. Zellforsch.*, **3**, 125–80.
Preer, J.B. and Preer, L.B. (1979). The size of macronuclear DNA and its relations to models for maintaining genetic balance. *J. Protozool.*, **26**, 14–8.
Prescott, D.M., Murti, K.G. and Bostock, C.J. (1973). Genetic apparatus of *Stylonichia* sp. *Nature*, **242**, 597–600.
Price, M.V. and Waser, N.M. (1982). Population structure, frequency-dependent selection and the maintenance of sexual reproduction. *Evolution*, **36**, 35–43.
Rae, P.M.M. and Spear, B.B. (1978). Macronuclear DNA of the hypotrichous ciliate *Oxytricha fallax*. *Proc. Nat. Acad. Sci. US*, **75**, 4992–6.
Raffel, D. (1930). The effect of conjugation within a clone of *Paramecium aurelia*. *Biol. Bull.*, **58**, 293–312.
Raikov, I.B. (1982). *The Protozoan Nucleus*. Transl. N. Bobrov and M. Verkhovtseva. Vienna: Springer–Verlag.
Rautmann, H. (1909). Der Einfluss der Temperatur auf das Grossenverhaltnis des Protoplasmakorpers zum Kern. Experimentelle Untersuchungen an *Paramecium caudatum*. *Arch. Zellforsch.*, **3**, 44–80.
Reisa, J.J. (1973). Ecology. In *Biology of Hydra*, ed. A.L. Burnett, pp. 59–105. New York: Academic Press.
Reynolds, C.S. (1984). *The Ecology of Freshwater Phytoplankton*. Cambridge Univ. Press.
Richards, O.W. and Dawson, J.A. (1927). The analysis of the division rates of ciliates. *J. gen. Physiol.*, **10**, 853–8.
Root, F.M. (1918). Inheritance in the asexual reproduction of *Centropyxis aculeata*. *Genetics*, **3**, 174–206.
Rose, M.R. (1983). Theories of life-history evolution. *Amer. Zool.*, **23**, 15–23.
Sacher, G.A. and Hart, R.W. (1971). Longevity, aging and comparative

cellular and molecular biology of the house mouse, *Mus musculus*, and the white-footed mouse, *Peromyscus leucopus*. *Birth Defects*, **14**, 71.
Schensted, I.V. (1958). Model of subnuclear segregation in the macronucleus of ciliates. *Amer. Natur.*, **92**, 161–70.
Schneider, E.L. and Mitsui, Y. (1976). The relationship between in vitro cellular aging and in vivo human age. *Proc. Nat. Acad. Sci. US*, **73**, 3584.
Schwartz, V. and Meister, H. (1975). Die Extinktion der feulgengefarbten Makronucleusanlage von *Paramecium bursaria* in der DNA-armen Phase. *Arch. Protistenk*, **117**, 60–4.
Seyffert, H.-M. (1979). Evidence for chromosomal macronuclear substructures in *Tetrahymena*. *J. Protozool.*, **26**, 66–74.
Siegel, R.W. (1970). Organellar damage and revision as a possible basis for intraclonal variation in *Paramecium*. *Genetics*, **66**, 305–14.
Shields, W.C. (1982). *Philopatry, Inbreeding and the Evolution of Sex*. Albany: State Univ. New York Press.
Simpson, S.B. and Cox, P.G. (1972). Studies on lizard myogenesis. In *Research in Muscle Development and the Muscle Spindle*. ed. B.G. Banker, R.J. Przybylski, J.P. van der Meulen and M. Vistor, p. 72. Amsterdam: Excerpta Medica.
Smith, J.R. and Hayflick, L. (1974). Variation in the life span of clones derived from human diploid cell strains. *J. Cell Biol.*, **62**, 48.
Smith-Sonneborn, J. (1971). Age-correlated sensitivity to ultraviolet radiation in *Paramecium*. *Genet. Res.*, **46**, 64–9.
Sonneborn, T.M. (1930). Genetic studies on *Stenostomum incaudatum* (nov sp). I. The nature and origin of differences among individuals formed during vegetative reproduction. *J. exp. Zool.*, **57**, 57–108.
Sonneborn, T.M. (1937). The extent of the interendomictic interval in *Paramecium aurelia* and some factors determining its variability. *J. exp. Zool.*, **75**, 471–502.
Sonneborn, T.M. (1938). The delayed occurrence and total omission of endomixis in selected lines of *Paramecium aurelia*. *Biol. Bull.*, **74**, 76–82.
Sonneborn, T.M. (1939). Genetic evidence of autogamy in *Paramecium aurelia*. *Anat. Rec.*, **75**, 85.
Sonneborn, T.M. (1947). Recent advances in the genetics of *Paramecium* and *Euplotes*. *Adv. Genetics*, **1**, 263–358.
Sonneborn, T.M. (1954). The relation of autogamy to senescence and rejuvenescence in *Paramecium aurelia*. *J. Protrozool.*, **1**, 38–53.
Sonneborn, T.M. and Lynch. R.S. (1937). Factors determining conjugation in *Paramecium aurelia*. III. A genetic factor: the origin at endomixis of genetic diversities. *Genetics*, **22**, 284–96.
Sonneborn, T.M. and Schneller, M. (1955). Age-induced mutations in *Paramecium*. *J. Protozool.* (*suppl.*), **2**, 6.
Sonneborn, T.M., Schneller, M. and Craig, M.F. (1956). *J. Protozool.* (*suppl.*) **3**, 8.
Spear, B.B. and Lauth, M.R. (1976). Polytene chromosomes of *Oxytricha*: biochemical and morphological changes during macronuclear development in a ciliated protozoan. *Chromosoma*, **54**, 1–13.

Spencer, H. (1924). Studies of a pedigree culture of *Paramecium calkinsi. J. Morphol.*, **39**, 543–51.
Spiegelman, S. (1971). An approach to the experimental analysis of precellular evolution. *Quart. Rev. Biophysics*, **4**, 213–53.
Steinbruck, G. (1986). Molecular reorganization during nuclear differentiation in ciliates. In *Germ Line-Soma Differentiation.* ed. W. Hennig, pp. 105–74. Berlin: Springer–Verlag.
Steinbruck, G., Haas, I., Hellmer, K.-H. and Ammermann, D. (1981). Characterization of macronuclear DNA in five species of ciliates. *Chromosoma*, **83**, 199–208.
Szilard, L. (1959). On the nature of the aging process. *Proc. Nat. Acad. Sci. US*, **45**, 30–45.
Threlkeld, S.T. (1976). Starvation and the size structure of zooplankton communities. *Freshwater Biology*, **6**, 489–96.
Tooby, J. (1982). Pathogens, polymorphism and the evolution of sex. *J. theoret. Biol.*, **97**, 557–76.
Turner, C.L. (1950). The reproductive potential of a single clone of *Pelmatohydra oligactis. Biol. Bull.*, **99**, 285–99.
Tyson, J.J. (1985). The coordination of cell growth and division – intentional or accidental? *BioEssays*, **2**, 72–7.
Uspenski and Uspenskaya. (1927). *Zeitschr. Bot.*, **17**, 273. (Not seen.)
Van't Hoff, J. and Sparrow, A.H. (1963). A relationship between DNA content, nuclear volume and minimum mitotic cycle time. *Proc. nat. Acad. Sci. US*, **49**, 897–902.
Vieweger, T. (1918). Les lignes continues des *Colpidium* Ehrbg. *C R Acad. Sci. Varsovie*, **11**, 611–22.
Wattiaux, J.M. (1968). Cumulative parental age effects in *Drosophila subobscura. Evolution*, **22**, 4306–421.
Weindruch, R.H. and Doerder, F.P. (1974). Age-dependent micronuclear deterioration in *Tetrahymena pyriformis*, syngen I., *Mech. Aging Devt.*, **4**, 263–79.
Weismann, A. (1889–1892). *Essays upon Heredity and Kindred Subjects.* vols. I and II. Oxford: Clarendon Press.
Whitney, D.D. (1912). Reinvigoration produced by cross fertilization in *Hydatina senta. J. exp. Zool.*, **11**, 339–59.
Wichterman, R. (1953). *The Biology of Paramecium.* New York: Blakiston.
Williams, D.B. (1980). Clonal aging in two species of *Spathidium* (Ciliophora: Gymnostomatida). *J. Protozool.*, **27**, 212–5.
Williams, D.B. and Williams, E.L. (1965). Aging and its relationship to ultraviolet light sensitivity in the ciliate *Spathidium spathula.* In *Progress in Protozoology.* ed. R.A. Neal, pp. 231–2. Amsterdam: *Excerpta Medica Found.*
Williams, G.C. (1957). Pleiotropy, natural selection and the evolution of senescence. *Evolution*, **11**, 398–411.
Williams, G.C. (1966). *Adaptation and Natural Selection: a critique of some current Evolutionary Thought.* Princeton, New Jersey: Princeton Univ. Press.
Williams, G.C. (1975). *Sex and Evolution.* Princeton New Jersey: Princeton Univ. Press.

Wolfe, M.S., Barrett, J.A. and Jenkins, J.E.E. (1981). The use of cultivar mixtures for disease control. In *Strategies for the Control of Cereal Diseases*. ed. J.F. Jenkyn and R.T. Plumb, pp. 73–80. Oxford: Blackwell.
Woodruff, L.L. (1905). An experimental study on the life history of hypotrichous infusoria. *J. exp. Zool.*, **2**, 585–632.
Woodruff, L.L. (1908). The life cycle of *Paramecium* when subjected to a varied environment. *Amer. Natur.*, **42**, 520–6.
Woodruff, L.L. (1909). Further studies on the life cycle of *Paramecium*. *Biol. Bull.*, **17**, 287–308.
Woodruff, L.L. (1911). Two thousand generations of *Paramecium*. *Arch. Protistenk.*, **21**, 263–6.
Woodruff, L.L. (1917a). Rhythms and endomixis in various races of *Paramecium aurelia*. *Biol. Bull.*, **33**, 51–56.
Woodruff, L.L. (1917b). The influence of general environmental conditions on the periodicity of endomixis in *Paramecium aurelia*. *Biol. Bull.*, **33**, 437–62.
Woodruff, L.L. (1926). Eleven thousand generations of *Paramecium*. *Quart. Rev. Biol.*, **1**, 436–8.
Woodruff, L.L. (1927). Studies on the life history of *Blepharisma undulans*. *Proc. Soc. exp. Biol. Med.*, **24**, 769–770.
Woodruff, L.L. (1928). *Paramecium aurelia* in pedigree culture for twenty-five years. *Trans. Amer. Microscop. Soc.*, **51**, 196–8.
Woodruff, L.L. and Baitsell, G.A. (1911a). The reproduction of *Paramecium aurelia* in a constant culture medium of beef extract. *J. exp. Zool.*, **11**, 135–42.
Woodruff, L.L. and Baitsell, G.A. (1911b). Rhythms in the reproductive activity of infusoria. *J. exp. Zool.*, **11**, 339–59.
Woodruff, L.L. and Erdmann, R. (1914). A normal periodic reorganization process without cell fusion in *Paramecium*. *J. exp. Zool.*, **17**, 425–518.
Woodruff, L.L. and Moore, E.L. (1924). On the longevity of *Spathidium spathula* without endomixis or conjugation. *Proc. Nat. Acad. Sci. US*, **10**, 183–6.
Woodruff, L.L. and Spencer, H. (1924). Studies on *Spathidium spathula*. II. The significance of conjugation. *J. exp. Zool.*, **39**, 133–96.
Wright, W.E. and Hayflick, L. (1975). Nuclear control of cell aging demonstrated by the hybridization of anucleate and whole cultured normal human fibroblasts. *Exp. Cell Res.*, **96**, 113.

Index of first authors

Index of genera

Index of subjects